HARVARD BOOKS IN BIOLOGY

Number 4: Nerve Cells and Insect Behavior

A female nymph of the praying mantis, *Hierodula sp.* Its posture suggests readiness for a predatory strike, but also contains a strong element of defensive display. This is shown by the slightly separated forelegs and the backward directed antennae.

Nerve Cells and Insect Behavior

REVISED EDITION

Kenneth D. Roeder, tufts university

Harvard University Press, Cambridge, Massachusetts

Second Printing, 1969

Distributed in Great Britain by Oxford University Press, London

---

Library of Congress Catalog Card Number 67-27092
SBN 674-60800-3

Printed in the United States of America

Book Design by David Ford

Preface to Revised Edition

The objective of this book remains unchanged. It is to record in readable form a rather personal exploration of some of the nervous mechanisms responsible for insect behavior. As such, the story is inevitably one-sided. Many exciting discoveries concerned with neurophysiology and learning have been neglected, particularly those at the cellular and biochemical level. This was done not because they are unimportant or because I am unaware of them, but because they do not now seem to be immediately relevant to the story I am trying to tell. The excitement of science is that these more distant discoveries may become immensely important to ones own field at any moment. But that is for future editions.

The main changes in this edition include the insertion of Chapters 6 and 11. These recount recent developments in analyzing the behavior of moths when they hear the cries of bats, and in tracking nerve signals from their ears through their central nervous systems. Progress in these directions has been due in large measure to my freedom from many academic responsibilities that was made possible by a Research Career Award from the National Institutes of Health. To them my heartfelt thanks.

February 1967 Kenneth D. Roeder

Preface

This book is an account of some aspects of nerve activity and insect behavior that have been of particular interest to me. It is not intended as a review of insect nerve physiology. Most of the chapters have been written as self-contained essays, just as I first explored each topic for its intrinsic interest rather than because it fitted into a major plan. But in reviewing these adventures I have also searched for a common connecting theme, and I have included some neurophysiological background in the hope of relating them to the mainstream of research on the neural basis of behavior.

The nature and direction of the book are perhaps best revealed by telling briefly how I came to be involved with insects, nerves, and behavior. An interest in insects began in early childhood, when my father introduced me to the excitement of collecting and rearing British butterflies and moths. Any child who has raised caterpillars and watched the extraordinary transformation of form and function during insect metamorphosis has been exposed to some of the central questions of animal behavior and development. I don't believe that I thought seriously about insects' nervous systems until I was a member of the physiology course given during the summer of 1932 at the Woods Hole Marine Biological Laboratory, when Dr. C. Ladd Prosser, at that time an instructor in the course, showed me the possibilities of experimenting on the arthropod central nervous system. Later, I was carried in an electrophysiological direction through the interest of Dr. Leonard Carmichael, then president of Tufts University, and through his generosity in allowing me to use the electrophysiological laboratory that he rarely had time to enjoy himself. In recent years the study of animal behavior has exerted a renewed pull on me through contact with Dr. Konrad Lorenz and many other ethologists. A number of their concepts, arrived at through long observation of animals under natural conditions, seem to me to have significant analogies in neurophysiological processes. In a sense, this book is a resifting of my own scientific experiences in order to find out whether any of these analogies are signs of deeper causal relations.

Any attempt to spin connecting threads between established and internally coherent fields of knowledge must inevitably distort both fields in the eyes of their orthodox adherents. The threads can be made to connect only at the peripheries of each field, so that the edges tend to become distorted out of proportion to the central core of each. The value of the threads and distortions lies in the transitions and new viewpoints that they uncover. I feel that insects provide an important transition point from ethology to neurophysiology, even though many ethologists probably consider that insects are dull subjects compared with fishes and birds, and most neurophysiologists will rightly claim that insects have not contributed to neurophysiology in the same degree as squid, frogs, and cats. However, if this book provides ethologists with a readable account of some of the notions of neurophysiology that pertain to behavior, and if it encourages some of the neurophysiologists to take time out from the oscilloscope screen for a little patient and passive observation of the normal behavior of their subjects, then it will have been worth while writing.

ACKNOWLEDGMENTS

It is impossible to name all the students, research associates, visitors, and colleagues who have influenced the thoughts and experiments on insects described in the following pages. A great many people have been and are actively associated with the work on insects in our laboratories at Tufts University, and I shall try to mention a few of these as they are connected with the topics of this book.

The experiments on ultrasonic hearing in moths (Chapters 4 and 5) began in 1955 when Dr. Asher Treat (The City College of New York) visited the laboratory in the hope that it might be possible to record nerve impulses from a moth's ear. His persistence in the face of my initial pessimism established a mutual feedback between us that still continues in the business of bats and moths. The experience of Dr. Donald Griffin (Harvard University) with bats, his encouragement from the sidelines, and his generosity in lending special equipment did much to aid this project.

The experiments of Chapter 7, also some of those in Chapters 9 and 10, are a few of the many that we have carried out at the expense of that alert, elegant, and most misunderstood of insects, *Periplaneta americana* L. The close association of Miss Elizabeth A. Weiant with our work on this insect deserves special mention, not only for her skill in fine surgery, but also for her perseverance in the face of initial antipathy. Others who played an active part in our work with cockroaches are Mrs. John L. Kennedy, Dr. Chester C. Roys, Mrs. Daniel Samson, Dr. Alan Slocombe, and Dr. Betty Twarog. In addition, the outstanding work of Dr. Phillip Ruck on photoreception and Dr. Nancy Milburn on neurosecretion is discussed in Chapters 9 and 10.

All of these are intramural acknowledgments. It is not feasible to go further afield since another book would be needed to discuss all the important work on insect sense organs, neuropharmacology and neurochemistry, orientation, and behavior that is entirely relevant to the matters to be discussed, but that has been omitted because a line had to be drawn somewhere.

Adequate funds for research are valuable for obvious reasons, but they also play an equally important part as a token of confidence. Twenty years ago a modest grant from the American Academy of Arts and Sciences played the latter role out of all proportion to its monetary value. Since then, the work in our laboratory has owed much to continuous and adequate support from contracts and grants from the Chemical Corps, United States Army, the National Science Foundation, and the United States Public Health Service.

I am grateful to the following publishers for permission to use some of the figures, in their original or modified form: John Wiley and Sons, Inc., New York (Fig. 32); The Johns Hopkins Press, Baltimore (Figs. 34 and 36); W. B. Saunders Company, Philadelphia (Fig. 37); University of Oregon Publications, Eugene, Oregon (Fig. 62); and Annual Reviews Incorporated, Palo Alto, California (Fig. 63). Previously unpublished figures were kindly supplied by Dr. J. J. G. McCue, Lincoln Laboratory, Massachusetts Institute of Technology (Fig. 6), Mr. Frederick A. Webster, Cambridge, Massachusetts (Figs. 16 and 19), and Dr. Nancy Milburn, Tufts University (Figs. 46 and 48).

The burden of typing the manuscript was born effectively by Mrs. Frances French, and the final drafting and lettering of many of the figures was carried out by Mr. George Johnson.

The book would probably never have been written if the Trustees of Tufts College had not granted me a leave from the information overload of present-day academic life, and if the Weiberhof had not provided hospitality, peace, and freedom from other responsibilities.

K. D. R.

Contents

Figures

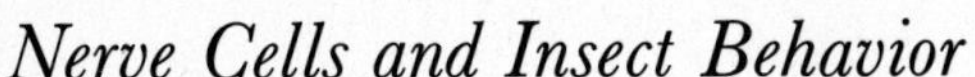

Nerve Cells and Insect Behavior

1. *Coding and Complexity*

Philosophers and scientists have invoked many special qualities in attempting to distinguish living from nonliving matter. However, as we learn more about the ultrastructure of proteins and viruses, chemical bonding, and the energetics and kinetics of living and nonliving systems, a sharp dividing line becomes harder and harder to establish. Living and nonliving systems are clearly composed of the same ingredients and subject to the same physical laws.

If life is not made of special stuff, then its uniqueness must lie in the pattern of its structural organization or molecular order. The secret is in the system and not in the materials. Indeed, if this were not the case it would be hard to explain the familiar cycle of growth and assimilation followed by death and decay.

One of the many attributes of living matter is its inevitable tendency to grow, differentiate, and become more complex. In living systems simple chemical compounds are elaborated into proteins, proteins into cells, cells into organisms, each stage requiring the synthesis of molecular configurations that are thermodynamically more improbable and unstable than those of the preceding stage. At first glance, these processes in living matter seem to contradict the second law of thermodynamics. Energy is constantly being transformed from one kind to another; thus, electric energy can be transformed into heat or into energy of motion or in other ways; mechanical or chemical energy can be transformed into electrical energy; heat energy can be transformed into mechanical energy, and so on. But in all of these changes some energy is transformed into heat and some of the heat descends to the temperature of the surroundings and thus becomes unavailable for further transformation into useful work. Thus in a closed universe (one into which energy cannot enter) the prospect is eventual transformation of all energy into heat and the degradation of this energy to a state in which none of it is available for transformation into any other form. This trend would be expected to reduce matter to its simplest possible form.

The apparent contradiction between this consequence of the second law of thermodynamics and the energy transformations that occur in living matter is reconciled when we realize that living matter consists of a multitude of open systems. Energy, in the form of sunlight or entrapped in the chemical bonds of nutrient compounds, streams continuously through each such system. Some of this energy is used to build more complex configurations involving new chemical bonds; some becomes unavailable by transformation into heat but there is a new supply constantly coming into the system from outside. The role of living matter may be likened to that of a hydraulic ram that employs some of the kinetic energy released by water flowing downhill to pump a small portion of the water uphill while the rest of it becomes unavailable for use. Thus the presence of life in one portion of the universe, taken as a closed system, need not alter the over-all process of degradation of energy in the universe, and the physicist is satisfied.

But the biologist is not. Most biological research is directly or indirectly concerned with reaching a fuller understanding of this synthesizing or organizing property of living matter. A penetrating discussion of living matter and the second law of thermodynamics is to be found in Blum's *Time's Arrow and Evolution,*[3] and the huge literature of genetics, systematics, evolution, embryology, morphology, biochemistry, and physiology deals with various facets. Although we are still unable to define with precision the difference between the living and the nonliving, these sciences are informing us about many of the associative properties of life.

The inevitable energy gradient in time predicted by the second law of thermodynamics is, provided we regard it statistically and without regard for the movements of individual atoms, essentially smooth and stepless. Events such as cosmic explosions are minor perturbations in the total energy flux—exceptions rather than the rule. The pathway of life toward order is likewise thought to be unmarked by any major steps or discontinuities. Whether we consider organic evolution from the earliest forms of life to man and his social systems, or the development of an egg into an adult, the picture is one of a continuous and multidirectional unfolding accompanied by many changes of pace and direction but never interrupted by major quantal jumps or saltations. There is as little scientific evidence for the special creation of an Adam and Eve as there is for special annihilation—the instantaneous disintegration and disappearance of a living organism. Even at death, when biological time stops and the second law of thermodynamics operates without the biological countercurrent, the transformation from the living to the nonliving is more gradual than is generally supposed.

The transformation promoted by living matter from simple to higher orders of organization requires a system for the storage and transmission of information from one stage to the next. Biological progress toward complexity would not be possible in the face of the opposing tendency toward randomness unless each stage profited in some way from the experience of those gone before. Perhaps the connection between information transmission and the synthesizing attribute of living matter is not immediately obvious, so let us consider a few examples.

Darwin recognized two important prerequisites when formulating his theory of evolution through natural selection. These are that like tends to beget like and that no two individuals of a species are identical. The biological basis of these conditions did not begin to be understood until the advent of modern genetics and cytology some fifty years after the publication of the *Origin of Species*. We now know that inheritable material in the form of genes is contributed to the offspring by both parents. The identity of each gene is determined in the chromosomes by a relatively simple chemical arrangement which we are now close to understanding. The identity and order of the genes within the chromosomes provide a template from which specific proteins are replicated. The nature and arrangement of proteins in the body determine the anatomic form and physiologic activity characteristic of the individual. Thus, the gene arrangement of an individual could be thought of as a compact code or formula for his body form.

In the formation of germ cells the gene code is replicated, and on fertilization the gene codes of both parents are combined. However, the replication process is not quite perfect, and the combination of parental gene codes does not result in a smooth blending. If this were the case, the outcome could only be uniformity and it would be impossible to account for the novelty and variety so evident in evolution. All organisms would be identical. This is avoided by two contrasting properties of great importance. First, individual genes may retain their identity and potency through many generations even when their capacity to express themselves in body form is suppressed by other genes. Unlike most biological entities, the genes have stability in time. Second, in the process of gene replication some genes may be omitted from the germ-cell nucleus, the gene order may be changed, or certain genes may lose their stability. These mutations or "copying errors" in the replication process probably occur by chance, that is, not in response to need or to the action of specific aspects of the environment, although the frequency of their occurrence may be altered by outside influences such as high-energy radiation. Once they have taken place, these genetic changes become incorporated in the replicated product and are thus transmitted to future generations.

Although this is clearly an example of the transmission of racial information in coded form from one generation to the next, it might be thought that mutations and sexual recombination would tend to garble the message and lead to degeneration and randomness. However, it is just these "errors" and alterations in the coded message that permit evolutionary progress. The ordering factor is the environment to which all the members of a population are exposed. The environment acts like a sieve that favors survival and reproduction of those variants or mutants having characteristics more advantageous than the norm of the population. Similarly, disadvantageous mutants (probably the great percentage) are prevented by death from contributing their special gene arrangements to the racial pool. The same type of variant probably reappears many times in a population, but the alteration in the gene code that produced it will become widely diffused through the population only when the environment alters in such a way that the bodily expression of this gene arrangement confers some advantage in survival and reproduction.

We can conclude from this digression into evolutionary theory that the mechanism for the origin of species requires (1) a fairly stable means for encoding, transmitting, and decoding bodily characteristics between one generation and the next, (2) the possibility of alterations in the coded message through mutation and sexual recombination, and (3) a means for selecting for transmission those alterations that favor survival.

It is not difficult to recognize an analogous situation in the evolution of human society. The ideas and actions of outstanding individuals are encoded in language, stored and transmitted in various media such as books, and decoded eventually into ideas and actions by other individuals. Alterations in the message take place when the thoughts and actions of the reader are influenced by what he reads, in other words, in the educational process. Society plays the part of the selective mechanism that determines which thoughts and actions are "good" and shall be encoded and transmitted to future generations. As in organic evolution, the selective sieve—society in one case and environment in the other—is an averaging device. Many mutants are not advantageous under present conditions and many ideas are ahead of their time. Both may persist because of the relative stability of the coded message, or they may recur *de novo* a number of times before the state of the sieve is such that they become advantageous.

A third example of the relation between information coding and the development of complex and nonrandom biological systems is to be found in the nervous system of many-celled animals. The following chapters will be concerned with the relation between nerve-impulse coding and insect behavior. As an introduction we will take a cursory glance at the origins and evolution of the nervous system.

Unfortunately we know next to nothing about the internal mechanisms that enable the Protozoa (single-celled animals) to move toward places that are favorable for survival and away from noxious areas. Although these reactions or taxes are frequently simple trial-and-error processes rather than movement along a gradient, they bring their possessors into optimal conditions of light and temperature, and into favorable concentrations of oxygen, carbon dioxide, and other chemicals, and differ little from taxes shown by many-celled animals with nervous systems. Free-swimming protozoa may be observed to collect in a dense cloud at one point. If the optimum condition is moved to another point the cloud migrates in a manner that reminds one of a regiment of soldiers obeying a command.

Higher organisms are thought to have evolved from single-celled ancestors in which the daughter cells budding off at cell division remained adhering to one another instead of swimming off separately. If we assume that the cells composing these ancestral metazoa (many-celled animals) had characteristics like those of present-day protozoa, we can visualize a mat or ball of cells moving through the water to optimal conditions, this time like a squad of soldiers or a chorus line in which the members are physically linked together or follow prearranged rules. Like the squad or chorus, this primordial metazoan presumably had a repertoire of maneuvers each of which could be equally well executed by each one of the members if detached from the group. Orientation of the colony as a unit would be possible only if each motile member had the same set of built-in responses and all members were exposed to the same stimulus, although not necessarily at the same intensity. Good performance by a chorus line requires that all members have the same training and that the whole chorus be exposed to the same conductor or orchestra.

From this it follows that such systems will become unmaneuverable if the number of members becomes so large that different parts of the system are exposed to different stimuli. Similar confusion will result if members of the system differentiate from each other in such a way that they respond in conflicting ways to the same stimulus. The only alternatives under these circumstances are either to abandon active maneuver—the course taken by the higher plants—or to develop a system of internal communications and decision-making so that certain members are able to determine the responses of the rest in a given situation. In other words, differentiation, growth, and movement becomes possible only if information can be transferred from one part of the system to another.

In higher animals this transfer is accomplished by the endocrine and nervous systems. In the endocrine system the message is a simple one, consisting of a highly specific chemical (hormone) that is released by a special group of cells and diffuses indiscriminately throughout the body. The specificity of the message resides in the fact that, although all the cells are exposed to it, only certain cells are differentiated in such a way that their activity is altered by it. The effect may be likened to that of a fire whistle: everyone hears it but only the firemen respond to it. The messages of the endocrine system are usually slower to take effect and longer lasting in their action, but the endocrine system and the nervous system are closely interconnected and their actions grade one into the other.

The information conveyed by the nervous system is more complex and is transmitted along specific channels so that only certain target cells are exposed to it. An external change is coded by sense cells into a nerve-impulse sequence. This is transmitted along nerves into the central nervous system where it is combined with the activity of other cells that may register other external changes, the internal state of the organism, or the effects of previous activity. The resultant of this central integration is conveyed in a nerve-impulse sequence to glands and muscles which decode it once more into action by the organism on the external world.

These three examples suffice to illustrate the connection between the coding and storage of information and the trend of living matter toward complexity and order. They are summarized and compared below:

Process	*Code*	*Information coded*
Organic evolution	Gene arrangement	Adaptive or "good" variants
Social evolution	Language and writing	Adaptive or "good" ideas or actions
Integration of many cells for unified action	Nerve impulses, hormones	External stimuli, previous experience

Before taking a closer look at the last of these systems it seems worth while to ask one final question about the general relation between coding and the trend toward complexity. Why is the intermediate step of coding necessary? Before the advent of modern science this intermediate step was not appreciated. Certain microscopists of the eighteenth century believed that they could see within the head of a sperm the body of a homunculus or a miniature human complete in every detail. Development of the embryo was thought to be merely a matter of the growth of this preformed body, which was supposed to contain sperm within sperm *ad infinitum—et absurdum.* Another belief that is held by many even today is that language can be circumvented by direct thought transference or telepathy from one mind to another.

There is no scientific justification for either of these beliefs and they are mentioned merely because they both omit the intermediate step of coding, yet it is hard to find a general justification for the added complication of the encoding-decoding process that seems to be inevitable in the biological sequence. One possibility is that the encoding of those steps in that part of the biological process that must have a considerable degree of stability and permanence removes them in some degree from the remorseless passage of biological time. By being encoded they are relatively segregated from the continuous energy flow and transformation that is characteristic of life, and thus are not time-dependent to the same degree as most biological events. The same argument might be applied to the nervous system with respect to the storage of information that determines future behavior on the basis of past individual experience, although this function of the nervous system does not appear to be important during the early stages of its evolution. Another reason for coding may be the amplification that it makes possible. This will be discussed in later chapters.

In summary, mechanisms for the coding, transformation, and storage of information seem to be inevitably associated with the synthesizing or organizing characteristics of living matter. As organisms become more complex their information-handling systems follow suit. For this reason our knowledge of the operation of the human nervous system in terms of its nerve components is still limited to a few of the simpler reflex mechanisms in the spinal cord. The number of nerve cells in the human brain makes the problem of analyzing its operation harder than that of analyzing the social structure of New York in terms of the activities of each of its inhabitants. An appropriate first step in the social study would seem to be a pilot analysis of a small New England village. In later chapters we shall examine the insects as a much simpler group of animals with the object of finding out if it is possible to correlate the coded information transmitted by their nerves with the way they behave. But first we must glance at the ways in which the relation between nerve function and behavior are being studied at the present time.

2. *Methods of Studying Animal Behavior*

Recent advances in biochemical genetics suggest that it will soon be possible to induce direct changes in the gene code with sufficient control to produce new characteristics imprinted on the germ plasm and therefore part of the racial inheritance. This possibility of synthesis implies a knowledge of inheritance mechanisms that is far in advance of our present knowledge of behavior mechanisms. Here most workers are still occupied with methods and with the empirical assembly of data. From many directions workers are tunneling hopefully into the mountain, some with steam shovels and others with dental drills. Some travel blindly in a circle and come out close to their point of entrance; some connect, usually in a mismatched fashion, with the burrows of others. Some have chosen to disregard the random activities of their fellows and have worked out in a small region an elegant system of interconnecting tunnels of their own. Both the attraction and the confusion of this multitudinous excavation lie in the fact that none of the workers know precisely what they are looking for or what they are likely to find.

Some of the confusion in attempts to understand the mechanisms of animal behavior lies in the conflict between two methodological concepts, both of which are perfectly acceptable in themselves. Anyone who has tried to observe songbirds or other wild animals knows that he must remain as inconspicuous as possible if he is to see anything of interest. The behavior of animals is as characteristic of their species as is their form, and the significance of behavior and form can be recognized only in their natural context. Therefore, it is self-evident that the observer must decouple himself from his subject as much as possible, first, so that his presence does not cause the behavior to deviate from its natural pattern, and second, so that he himself is not seduced into anthropomorphic interpretations of what he sees. Of course, neither of these two aims can be fully realized, and any claims to complete objectivity in behavior studies must be regarded with suspicion.

The second methodological concept needs no elaboration. It is the well-known scientific principle that in a situation containing several variables the effect of each must be examined in turn while the others are kept constant.

No student of animal behavior would deny the correctness of both procedures, yet they imply experimental situations that contrast and sometimes conflict with one another. Should the animal be free in its natural environment so that it can show "normal" behavior, or should it be placed in an experimental chamber where environmental factors can be controlled? Should the animal or the observer be isolated from the external environment? Many of the misunderstandings and disjunctions of research efforts on animal behavior can be traced to a leaning toward either one or the other of these justifiable but contradictory procedures.

The ethologists[23,57,58] use methods of gathering information that depend primarily upon isolation of the observer, leaving the animal as free as possible in its natural environment. The observer may be represented as surrounding himself with a blind (Fig. 1*A*) from which he observes his subject with minimum interference in the natural situation. The

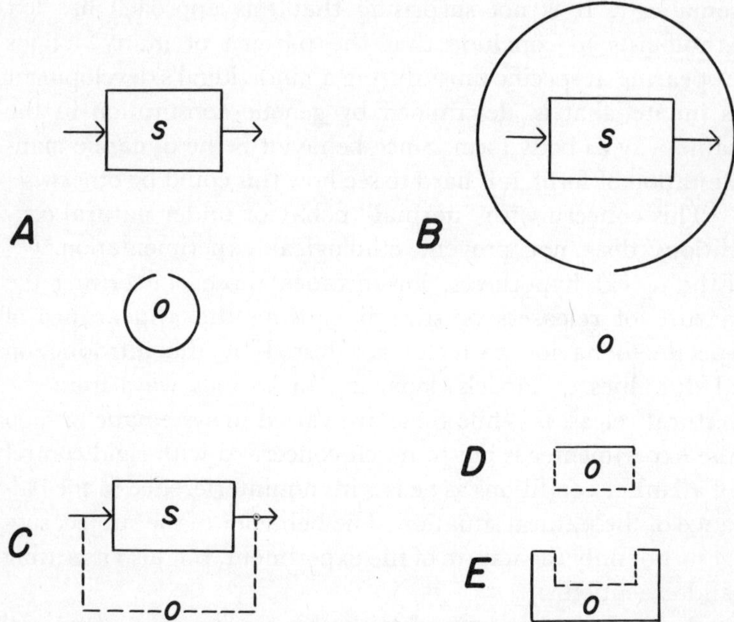

Fig. 1. Methods of studying animal behavior: (*A*) ethological approach; (*B*) approach of experimental psychology; (*C*) approach of students of animal orientation; (*D, E*) "single-unit" and "ablation" methods of physiology; *o*, observer; *s*, subject. The circle in (*A*) represents the blind surrounding the observer, and that in (*B*) the experimental chamber surrounding the animal.

animal remains as free as possible to act on, and be acted on by, its environment (arrows). After extensive observation of conspecific, interspecific, and environmental interactions of a species, the ethologist assembles an ethogram or "behavioral morphology" of his subject. Similar studies on related species and higher systematic categories lead to the generalizations of ethology. Concern with the adaptive value and phylogeny of behavior patterns, and with the signal function of certain ritualized actions in maintaining species isolation, places ethology in close relation to comparative anatomy and systematics[23]. It is not surprising that this approach has led ethologists to conclude that the pattern of many actions appearing at specific times during an individual's development is innate, that is, determined by genetic constitution in the same way as body form. Since behavior is the dynamic manifestation of form, it is hard to see how this could be otherwise.

This concern with "normal" behavior under natural conditions does not prevent ethological experimentation.[57,58] Ethological hypotheses, for instance those concerning the nature of releasers or stimuli causing the appearance of specific behavior patterns, are tested by the introduction of dummies or models departing in various ways from the natural releaser. While these are varied in systematic fashion the experimenter is not so much concerned with rigid control of all other conditions as he is with noninterference in the balance of the natural situation. The behavior of the subject suggests not only the nature of the experiment, but also its setting and execution.

Psychologists concerned with the mechanisms of animal behavior usually lean more toward the second experimental method, that is, isolation of the subject from its normal environment and from the observer by enclosing it in a cage or experimental chamber where conditions are controllable and may be varied one at a time. In short, the blind that surrounds the ethologist has now been turned around so that it encloses the animal (Fig. 1*B*).

The advantage of this method is the control it gives the experimenter over conditions, thereby greatly increasing the probability that his results can be confirmed by others. No scientist would question the importance of this, since experiments that cannot be repeated have no value. Yet, in behavior studies the desire to control all the conditions surrounding an observation or experiment can become an obsession that may rob the conclusions of any meaning except as a special case. The ethologists have been accused of making generalizations based on information bordering on the anecdotal, such as could be reobtained by another observer only with the greatest good fortune. Behavioral psychologists have been criticized for generalizations which, though readily confirmed, have been made under conditions so far removed from those surrounding wild members of the subject species that the behavior in question can be considered so pathological and so strongly tainted with human interference as to bear no relation to the behavior that brought the animal to its present evolutionary status. There seems to be no way of circumventing this methodological paradox, but the great interest shown by both groups of workers in what the other is doing is leading to active attempts to utilize what is good in the methods of both.

Somewhere intermediate between behavioral psychologists, ethologists, and physiologists are the students of animal orientation.[27] Though they are concerned only with segments of the total behavioral pattern—those steering the animal in its environment—they mainly base their deductions on observations and experiments carried out with whole animals.

Steering may be defined as the correction of random perturbations encountered by the animal as it moves in constant relation to a physical or chemical gradient. Deviations from a constant course may be corrected only after they have occurred, as in the maintenance of constant temperature by a thermostat, or they may be anticipated, as when one lights a fire after reading the weather forecast.

These corrective or anticipatory adjustments made by animals to environmental gradients and changes are examined by inserting the observer and his measuring instruments into the loop connecting the animal with its environment (Fig. 1*C*). To take a simple but common example, an animal may steer a straight course by moving so as to keep a constant angle with the rays of the sun.[12,22,27] Perturbations in the medium, such as irregularities in the ground or eddies in air or water, tend to alter this angle in a random fashion as the animal progresses. The animal reacts to this through negative feedback, that is, any tendency of the selected light angle to *increase* causes the animal to make movements that *reduce* it, and vice versa.

Hypothetical models of the mechanisms necessary for this and for much more complex reactions are constructed from observed changes in the reactions when the guiding gradient is manipulated artificially.[27] These models are derived without concern for the actual neural and motor mechanisms employed by the animal, but are in a sense predictions of the sort of mechanisms one might expect to find connecting the receptor and effector systems. Thus, they provide a powerful guide to physiological exploration (see Chapter 12).

Animals that steer their courses by sun, moon, or stars over long distances or for long periods require also a time sense or internal "clock" if they are to avoid traveling in loops.[2] This applies also to forms showing circadial rhythms, that is, cycles of activity about 24 hours in length and corresponding approximately to the alternation of day and night, that persist even when the subjects are isolated from the familiar signs of the earth's rotation. The study of "clocks" in animals and plants also requires insertion of the observer into the connection between the animal and its environment.

Students of animal orientation may carry out their experiments under controlled laboratory conditions, but their work begins and their objectives lie with the orientation of the animal under natural conditions. Since their primary method is to insert themselves into the feedback loop connecting the animal and its environment, and, since they are not immediately concerned with the actual mechanisms whereby it makes its adjustments, they work usually with intact animals.

However, their hypotheses and models are often tested and enlarged by experiments involving the removal or an alteration in the arrangement of sense organs and motor mechanisms of their subjects, which brings them close to the physiologists.

Physiological research on the relation of neural function to behavior is predominantly analytical in emphasis. Physiology has roots in anatomy and chemistry, where the first step is to subdivide complex organs and compounds into simpler components in the search for common denominators. In chemistry the ability to plan and execute a synthesis of a complex substance is a tribute to the information to be gained by the analytic method, but in anatomy and in neurophysiology the path still leads mostly in one direction.

As far as behavior mechanisms are concerned this analytic approach immediately leads the neurophysiologist from the intact animal to an examination of the component parts of the nervous system. This examination may be said to follow two directions: either the behavior of individual or grouped neurons, receptor cells, or muscle fibers is directly examined *in situ* or in isolation (Fig. 1*D*), or their function is indirectly inferred from changes in the behavior of the rest of the animal following their removal (Fig. 1*E*).

This emphasis on analysis is seen in the latest neurophysiological tools, most of which are directed at examining the activities of single cells. The application of some of these will be discussed more fully in the following chapters.

Even though it can be said from personal experience that preoccupation with the behavior of single neurons has a fascination of its own, analysis at a finer and finer level is not an end in itself as far as the topic of this book is concerned. There must be concern for the organization of the system even when the operation of its units is still incompletely understood. It seems almost as though analysis—taking apart—is a self-evident procedure to the human mind, whereas synthesis is not. Who does not recall that first thrill of analysis when the springs and gears of one's first watch were strewn upon the table, and the consternation when they could not be reassembled?

It has been the intention of this chapter to outline the general approaches taken by four different groups of workers toward an understanding of the mechanisms of animal behavior. If critical comments about the methods of one or the other group are read out of what has been said, it was certainly not my intent to suggest that success was more likely in one direction or the other. First, as was suggested in the simile of the mountain, no group has come even close to its objectives. Second, division into ethologists, psychologists, physiologists, and students of orientation was made purely for purposes of description, there being many workers whose methods and approach fall between two or more of these categories. In the chapters to follow much will be borrowed from each of these fields, and an attempt will be made to relate them in order to gain an insight into the mechanisms of insect behavior.

3. *Communication*

The coordinated actions of any system, whether it be the state or the association of cells that is a living organism, depend upon two general types of operation: discrimination and communication. Discrimination or decision-making is far more complex and far less understood than communication. Discrimination in the nervous system will be discussed in Chapter 8, and is mentioned here only to distinguish it from communication.

A sense cell discriminates one stimulus mode from all others when an impulse is generated in its nerve fiber only by an external change of a certain kind, such as light, or sound, or chemicals. It serves as a filter in distinguishing one stimulus mode from others, and as a transducer and amplifier by converting a minute energy change at its receptive surface into the entirely different and much larger energy change detectable as an impulse in its axon. The mystery lies not in the quantitative relation between stimulus and response, which has been worked out with some precision for many sense cells, but in the manner whereby the decision to respond is reached. Similar discriminations performed by neurons at all levels in the central nervous system (Chapter 8) are an important mechanism of behavior. It is also necessary to underline the importance of nonresponsiveness under certain conditions in sense cells, in central neurons, and in behavior itself.

Once a sense cell has responded, the mechanism whereby this decision is communicated to other nerve cells, and the form in which the neural message is coded, are not too difficult to understand. First, an attempt will be made to see whether what can be learned from "wire tapping" on nerve fibers communicating decisions in this network of discriminators bears any relation to behavior.

Animal behavior is based upon two well-known systems of internal communication, and others will probably be discovered. Chemically specific substances (hormones, neurohumors, mediators) are produced and released by specific cells, whence they diffuse or are transported over very short or often very great distances within the body, finally to produce specific responses in certain uniquely sensitive cells. Many cells are exposed but only few respond. On the other hand, the influence of nerve impulses during their transmission is limited to the surface of the transmitting nerve fiber. All nerve impulses appear to be due to the same type of physicochemical event, and their specificity in communication resides not in their nature, as is the case of the hormones, but in the pathways over which they travel.

Both systems of communication are closely interconnected, and frequently form alternate steps in the transmission of a biological message. The chemical steps always take longer to execute, and may range in duration from 0.5 millisecond in the case of neural mediators to days or weeks in the case of hormones.

The Nerve Fiber. Our concern is primarily with the behaviorally significant information transmitted by nerve impulses, and not with their physicochemical basis. However, it will be necessary to give a perfunctory description of the generation of nerve impulses in order to assess their potentialities as information carriers. Further details must be sought in books on neurophysiology.[5,10,52]

Each nerve fiber or axon is a thin-walled tube ranging in length from a millimeter to a meter or more and having a diameter of less than 1 to about 500 microns (abbreviated 500μ; 1000 microns = 1 millimeter). Human axons are 20 microns or less in diameter. Some of the axons found in the nervous system of insects are 50 microns in diameter but mostly they are smaller than this. Many worms and squid have some axons as large as 500 microns. Figure 2*A* shows the nerve cord of a cockroach in a cross section that contains all the axons connecting the fore and hind parts of the body. The membrane or wall of each axon appears as a darker oval enclosing the jellylike axoplasm. The axons are packed fairly

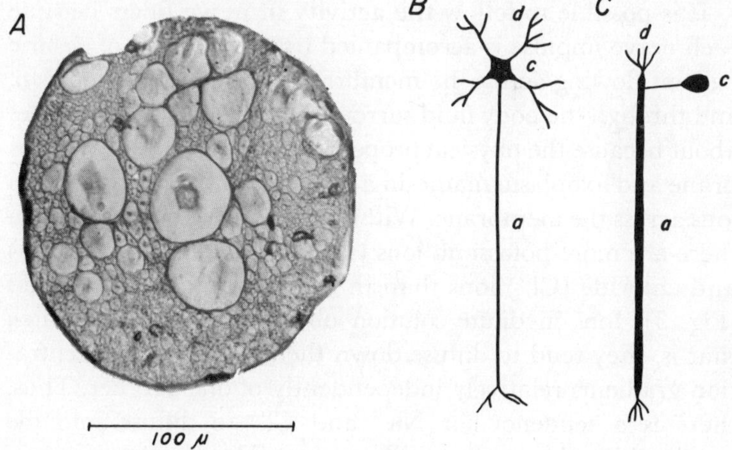

Fig. 2. (*A*) Cross section of one connective in the abdominal nerve cord of the cockroach, *Periplaneta americana*. (*B*) Bipolar neuron typically found in the central nervous system of vertebrates. (*C*) Unipolar neuron typically found in the central nervous systems of arthropods. *a,* axon; *c,* cell body or soma; *d,* dendrites.

closely, but between them can be seen the darker nuclei of supporting and ensheathing cells. The whole bundle of axons making up the nerve cord is enclosed in a thicker sheath. The two groups of larger axons visible near the center of the nerve cord will be discussed later.

If a single axon is followed as it runs among others through the nervous system it is seen to be a continuous tube varying in diameter but uninterrupted by partitions. It is usually found to be the greatly elongated outgrowth of a single nerve cell or neuron (Fig. 2), but some of the giant axons found in invertebrates originate from several cell bodies. If an axon is severed, the part separated from the cell bodies dies, while a new outgrowth is regenerated from the stump. This shows that the cell body plays a vital part in the maintenance of the axon, and that together axon and cell body constitute a single cell unit.

It is possible to follow the activity in nerve fibers because each nerve impulse is accompanied by a brief pulse of electric current flowing across the membrane, through the axoplasm, and through the body fluid surrounding the nerve. This comes about because the physical properties and metabolism of membrane and axoplasm maintain a constant imbalance of certain ions across the membrane. Within the inactive but living axon there are more potassium ions (K^+) and fewer sodium (Na^+) and chloride (Cl^-) ions than in the surrounding body fluid (Fig. 3). Ions in dilute solution obey the familiar gas laws, that is, they tend to diffuse down their respective concentration gradients relatively independently of one another. Thus, there is a tendency for Na^+ and Cl^- to diffuse into the axon, while K^+ tends to diffuse out. The apparent imbalance of ions normally found in the inactive axon is maintained by several processes. A passive though central part is played by an anion (A^-, Fig. 3) that is unable to diffuse outward through the membrane. It can be said to hold the excess of

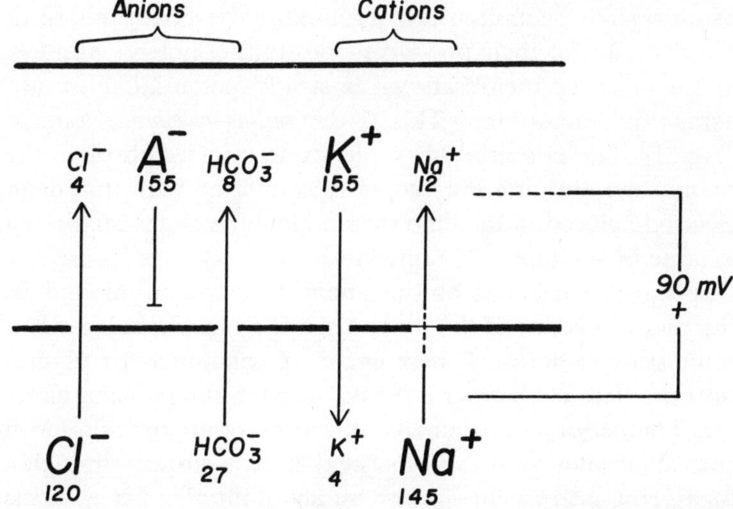

Fig. 3. The distribution of major ions between the interior of a vertebrate muscle fiber and the surrounding fluid. A similar situation exists at the surface of a nerve fiber. Numbers give concentrations (M/ml),[52] and show the direction and steepness of the diffusion gradient for each ion. A^- cannot pass through the living membrane. Na^+ cannot enter (or is continuously ejected) when the membrane is at rest (dashed line). This restriction is removed during activity, the Na^+ influx being responsible for the first phase of the action potential. K^+, Cl^-, and HCO_3^- have some freedom to move down their gradients ("pores" and arrows) in the resting state, but this tendency is self-limiting because it causes an excess of negative charges within (expressed by membrane potential of −90 mV) that counteracts further net movement.

K^+ within the axon by virtue of its charge. Other factors are less clear, but the dynamic equilibrium that exists is ultimately dependent upon continuous metabolic expenditure by the living nerve cell. It the axon is deprived of oxygen, or the membrane is destroyed, the ions diffuse down their respective concentration gradients until static equilibrium is reached.

In spite of the restraints to their net movements imposed by the living axon membrane, the ions exert steady diffusion pressures like the molecules of a gas. Since the ions are charged, the resultant of their pressures (mostly those of K^+ and Cl^- in the inactive membrane) is a steady potential difference across the membrane. This is the *resting-membrane potential* (Fig. 3). The potential of an electrode inserted through the membrane and into the axoplasm compared with that of an electrode placed in the fluid surrounding the axon usually has a value of −60 to −90 millivolts.

The measurement of this membrane potential and its fluctuations is one of the main tasks of neurophysiology. It is interesting to note that here again we encounter the ubiquitous problem of observer error discussed in the previous chapter. The act of measurement—insertion of an electrode with consequent flow of nerve-generated electric current through a measuring instrument—is also an act of interference with the source of what we are trying to measure. The situation is analogous to an attempt to determine the pressure in a cylinder supposed to contain gas. Whether we use the relatively crude measure of allowing a small current of gas to escape by opening the valve, or whether we attach an appropriate pressure meter, the gas pressure has been changed by our intervention. Most of the technical efforts in neurophysiology are directed to reducing this factor to a minimum by building smaller electrodes and electronic amplifiers requiring less current for their operation. Of course, the problem is a central one in most of science, particularly in physics. Perhaps it weighs less on biologists than on physicists since although their living subjects are heterogeneous and unpredictable their capacity for adjustment and self-repair offset the observer problem in some measure.

The Nature of the Nerve Impulse. The invasion of the axon membrane by a nerve impulse is accompanied by a brief but dramatic change in the dynamic balance of ions at the axon surface (Fig. 3). The capacity of the axon to extrude sodium ions is temporarily lost. Since the inside of the axon is negative and therefore deficient in positive charges, and there is an excess of Na^+ ions outside, everything is in favor of an inward rush of sodium ions when this membrane change takes place. This sudden influx of positive charges not only reduces the membrane potential to zero, but momentarily reverses it so that an electrode inside becomes positive compared with one placed in the surrounding body fluid (Fig. 4). A sodium influx is permitted for only 0.5 millisecond or less, at the end of which time the membrane begins to regain its resting property of excluding or extruding sodium ions. At this time movements of chloride inward and of potassium outward restore the status quo, and the membrane resumes its resting condition and its resting potential. The brief electrical change

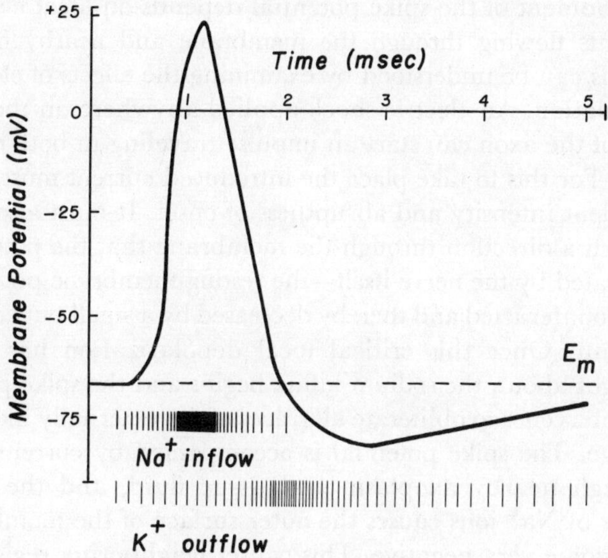

Fig. 4. The change in membrane potential registered by an intracellular electrode during a nerve impulse. The rising phase of the spike potential is due to inward movement of Na^+; the falling phase and eventual return to the resting-membrane potential level E_m is brought about largely by increased outward movement of K^+.

signaling these ion movements is known as the *action potential.* Its major portion or *spike* lasts for 0.5 to several milliseconds in different nerve fibers, and its voltage change is about 120 millivolts. The sodium ions accumulated by the axon during the spike potential are eventually pumped out in exchange for potassium ions with no significant potential change.

It is easier to understand how the nerve impulse is transmitted throughout the neuron if it is related to a general class of phenomena usually designated regenerative or self-propagating. These are chain reactions and employ positive feedback. A nuclear explosion, a burning fuze, and an avalanche are roughly analogous to a transmitted nerve impluse. On the other hand, the familiar vehicle of man-made communication, electric current in a wire, is not analogous even though the development of the spike potential depends on local electric currents flowing through the membrane and nearby fluids.

This can be understood by examining the effects of electric stimulation. An electric shock applied anywhere on the surface of the axon can start an impulse traveling in both directions. For this to take place the introduced current must have sufficient intensity and abruptness of onset. It must also flow in such a direction through the membrane that the potential generated by the nerve itself—the resting membrane potential—is counteracted and thereby decreased by a small but critical amount. Once this critical local depolarization has been brought about, the sodium influx begins and the spike potential intervenes to obliterate all traces of the externally induced change. The spike potential is accompanied by current flow through nearby axoplasm and tissue fluid, and the local influx of Na^+ ions causes the outer surface of the membrane to become very negative. This causes neighboring regions of resting axon membrane to become negative to some degree. When negativity in adjacent regions reaches a critical value a spike potential flares up, and the process propagates steadily throughout the length of the axon.

The difference between this process and the passage of electric current in a wire is that the metabolism of the living axon makes the impulse possible by maintaining the ion imbalance over the length of its surface; the wire merely offers resistance to the current arising from an ion imbalance in a battery connected to its ends. The metabolism of the axon can be likened to the deposition of snow on the top of a mountain by solar energy. Any disturbance causing a little snow to roll down hill can set off an avalanche that carries all the snow on the mountainside into the valley. Movement at the edge of this avalanche can trigger another in a neighboring valley, and so on. Then there must be further steady expenditure of solar energy to store water in this improbable spot before another avalanche can occur. In nerve, gunpowder, and avalanche, energy of chemical or nuclear origin is stored as energy of position in ions, atoms, or water. These are so poised that when a few units are disturbed part of the kinetic energy released disturbs a larger number and a chain reaction takes place.

A little reflection on the kind of system revealed by these imperfect analogies shows that certain other characteristics are inevitable. First, the propagated process is all-or-none, that is, it must go to completion. This does not mean that the action potential is the same size everywhere along the axon. Its size at any point will depend always upon local conditions such as the size of the membrane potential. Second, there may be all sorts of local changes that are reversible, provided they never sum in time or place to the critical threshold magnitude necessary to set off the propagated process. Third, once started, the propagated process bears no relation in size or speed to the stimulus that started it. Fourth, the threshold local disturbance can be produced in many ways; mechanical, chemical, osmotic, and electric stimuli can start identical nerve impulses. Fifth, the propagated impulse is invariably a transient rather than a continuous event. There must be a period of nonexcitability (refractory period) during which metabolic processes in the fiber restore ion gradients across the membrane to their resting levels. During the early stages, at least, of this period of "reloading" the system must be completely immune to external stimuli if it is to exhibit all-or-none behavior.

Several factors play a part in determining the velocity, spike duration, and refractory period of the nerve impulse. Increased temperature shortens all of these time functions, as it does in the case of other biological events. Greater axon diameter and the presence of a myelin sheath (as in many vertebrate axons) are associated with greater conduction velocity. The time course of local electrical responses of the membrane and of the propagated impulse is determined by the resistance to ion movement of the local electric circuits formed by the membrane and tissue fluids, and also by the capacitance or the quantity of charge (number of ions) that must be transferred across the membrane to cause a given change in its potential.

The velocity of impulse transmission in various nerves ranges from a few centimeters per second through 10 to 50 meters per second in the giant axons found in many invertebrates up to over 100 meters per second for the large myelinated axons of mammals. Spike potentials have durations of less than 1 millisecond to several milliseconds, and refractory periods have roughly similar values. The maximum number of impulses that can be transmitted by a single fiber in a second ranges up to somewhat over 1000 although many fibers have an upper limit that is much less than this.

Detection of the Action Potential. The transient change in membrane potential that signals the invasion of a region of axon by an impulse can be examined at full amplitude and with minimum distortion only by intracellular recording. In this method, already mentioned above, a glass pipette with an open tip 0.5 micron in diameter and filled with saline solution is introduced into the axoplasm and its potential is measured in relation to that of another electrode in the saline solution or body fluid outside. This registers the spike in the form shown in Figs. 4 and 5*C*. Intracellular recording is being

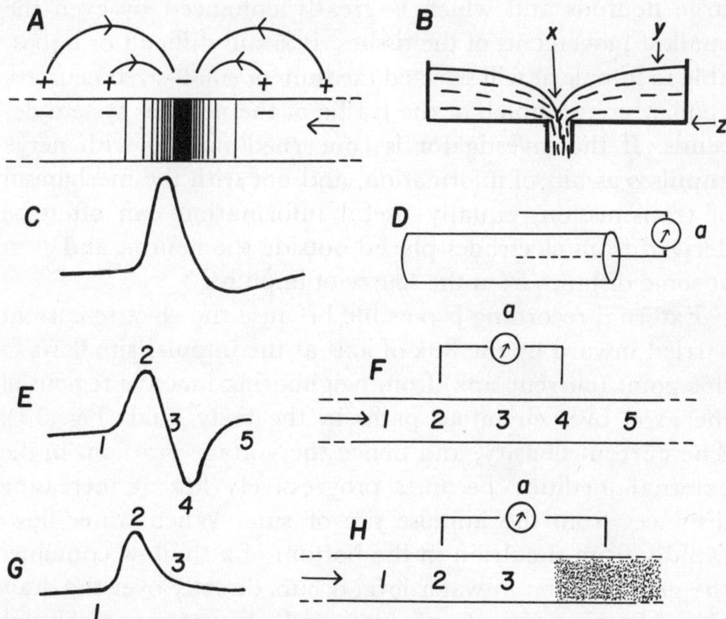

Fig. 5. Methods of recording action potentials. (*A*) The electrically negative "sink" in the region of a nerve momentarily occupied by an action potential. Current flows externally from inactive regions toward the sink. (*B*) Hydraulic analogy; the greatest depression of water level occurs directly over the sink. (*C*) Action potential recorded by intracellular arrangement shown in (*D*) (see also Fig. 4); it corresponds to water-level changes recorded between *x* and *z*. (*E*) Diphasic action potential registered by extracellular-electrode arrangement shown in (*F*); it corresponds to water-level changes recorded between *x* and *y*. (*G*) Monophasic action potential recorded by extracellular arrangement shown in (*H*), where one electrode is on crushed or inactivated part of nerve; it corresponds to water-level changes recorded between *y* and *z*. *a*, amplifying and recording apparatus; numbers indicate successive positions of the action potential responsible for corresponding sections of the recording.

more widely used, even with neurons lying deep within the central nervous system (Chapter 8). Its advantages are that the intimate activity of the impaled neuron can be isolated from that of its neighbors. Its disadvantages include the damage done by inserting the electrode, which limits its use to large neurons and which is greatly enhanced by even the smallest movements of the tissues. It is still difficult or impossible to impale at will selected medium or small-sized neurons, upon which so much of the traffic of the nervous system depends. If the investigator is concerned mainly with nerve impulses as bits of information, and not with the mechanism of transmission, equally useful information can often be derived from electrodes placed outside the neuron and even at some distance from the source of impulses.

External recording is possible because the electric current carried inward by the flux of ions at the impulse site flows to this point (current sink) from neighboring inactive regions of the axon over circuitous paths in the body fluid (Fig. 5*A*). The current density, and hence the voltage gradient in the external medium, becomes progressively less at increasing distances from the impulse site or sink. When water flows rapidly from the drain in the bottom of a shallow container the greatest drop in water level occurs directly over the drain (Fig. 5*B*). However, the presence and whereabouts of this sink are detectable at a distance as a slight drop in the water level (equivalent to voltage change in the nerve) due to the current flowing toward the sink. With this analogy in mind it will be seen that a measurement of the voltage difference between two electrodes (analogous to water-level difference) placed in a plane parallel to the direction of current flow will measure, admittedly with some degree of distortion, changes in the rate of flow of current at the active site (analogous to the flow of water through the drain opening). The shorter the distance separating the members of the electrode pair, or the greater the distance separating them from the source of current flow, the greater must be the sensitivity of the measuring device (Fig. 5*E, F*).

This is about as far as the hydraulic analogy can safely carry us. It can be patched up by constructing an interesting piece of plumbing in which the drain moves along the bottom of the tub to produce a traveling sink, while a pump simultaneously returns the draining water through a series of holes in the side of the tub. But the analogy strains the imagination almost as much as the reality.

In one method of external recording both electrodes are placed at different points on the outside of a nerve and connected to an amplifier that responds to the potential difference between them. A cathode-ray oscilloscope or other device displays the output of the amplifier as a time function. The traveling negative sink created by the propagated impulse passes closer first to one and then to the other of the electrode pair, causing a diphasic fluctuation to appear on the oscilloscope (Fig. 5*E*). If one of the electrodes is nearer to the nerve, and hence more influenced by the traveling sink than the other, the passage of the impulse past the near electrode will be registered as a monophasic curve whose form roughly approximates to the current flow in time across the active membrane nearest the electrode (Fig. 5*G, H*). The voltage change is, of course, very much less than that occurring across the membrane in the region of the sink.

Extracellular recording, though easier to apply and less damaging to the living neuron, also has serious drawbacks. Most nerves and nerve tracts are made up of a heterogeneous population of axons of all sizes and functions. Sensory, motor, and internuncial (connecting sensory and motor units) fibers are only rarely arranged in conveniently separated bundles. An electrode pair inserted into the fluids surrounding such a collection of nerve fibers encounters a welter of spikes of all sizes due to impulses passing in either direction in nearby or in more distant axons. Here the bathtub analogy becomes quite useless, for the tub must be equipped with many independently moving outlets that are to be distinguished one from another by changes in the water level.

There are several ways of achieving partial discrimination of impulses being transmitted along specific axons from the "noise" of the general traffic in a large nerve or tract. The best method is intracellular recording, but this is rarely practicable for the reasons given above. An extracellular electrode may be inserted blindly in the hope of getting close to an individual axon, but this is like hunting for a keyhole in the dark. Unwanted activity may be reduced by shielding the majority of neurons from their normal sources of stimulation, or by cutting all except a few fibers before they reach the recording electrode. Functional isolation of a given neuron may be obtained by applying a periodic stimulus to its ending and searching in the welter of action potentials for a spike that recurs regularly at the stimulus frequency. If the axon in question is larger than others in a bundle its impulse produces a proportionally greater current in the region of a nearby electrode, even though the voltage change at the sink is the same. Therefore, its activity can often be distinguished from the rest because its spike potential will appear to the electrode to be larger than those generated by surrounding fibers of smaller diameter. The extracellular method is discussed at greater length in Chapter 11, where it is used in tracing interneurons in a moth's central nervous system.

A difficulty of another type is encountered when the objective is to relate the information transmitted as spike activity from one part of the nervous system to another with the behavior of the whole animal. Hundreds of thousands of sense cells are arrayed in each of the major sensory fields—vision, hearing, and so on—in the higher vertebrates. Presentation of a behaviorally significant pattern or tone to the appropriate field probably excites or depresses the activity in some considerable fraction of the sense cells in the array. However, the significance of the stimulus must reside in the pattern or sequence of sense cells affected, and not in their number. Each sense cell directly or indirectly communicates its private decision as some change in the frequency of spike potentials in an afferent axon. It is hard to see how registration of this spike sequence in one or a few of these afferent axons among the many thousands active could reveal any relation between the total sensory message and the behavior. One might as well expect to derive a detailed statement of national policy from an interview with one citizen selected at random. The chance of picking the Secretary of State would be remote.

This problem of the manner in which sense organs communicate the uniqueness of a particular pattern, musical chord, odor, or taste over multiple parallel nerve fibers to the central nervous system is being studied intensively in the higher animals. A less ambitious approach is to choose a subject having a sense organ connected to the nervous system by a small number of nerve fibers. We cannot expect that the world surveyed by such a sense organ will be as full of nuances as our own, but it does give us the opportunity to apply available methods of neurophysiology to a problem in behavior.

4. *The Tympanic-nerve Response in Noctuid Moths*

The ear of a moth may seem to be a somewhat esoteric subject for a study of the form in which information is coded in nerve impulses. However, it has two outstanding virtues. First, in some families of moths, notably the owlet moths or Noctuidae, the tympanic organ contains only two acoustic sense cells. Electrodes placed on the tympanic nerve containing the axons from these sense cells can intercept all of the impulse-coded information this sense organ is capable of delivering to the moth's central nervous system. It is relatively easy to distinguish and interpret the spike patterns belonging to these two fibers, and hence to have at hand all the information reaching the central nervous system of the moth by this channel.[47,48,49] The second virtue is that the behavior of the acoustic receptors is quite conventional. The relation between stimulus and response is similar to that found by other workers in the nerves of other animals, though their methods often required the isolation of single units from among many thousands. The moth tympanic organ should provide a useful link between neurophysiology and behavior.

Even though this alone might seem ample justification for proceeding with the analytic task of physiology, it would have little meaning to the ethologist unless the inquiry is related to the pressures of natural selection responsible for the evolution of a sense of hearing in moths.

The selection pressure is not hard to find, since it is connected in a most direct manner with racial survival. Moths are one of the main food sources of certain families of bats. They are attacked on the wing and in darkness in a contest in which speed and maneuverability are the premium qualities. The fact that the contest has continued probably for some millions of years tells us that it is a balanced "game" played in the dark: all bats locate and capture some moths; some moths locate and evade all bats.

Here the link between predator and prey is acoustic. While in flight, insectivorous bats emit a series of brief chirps pitched several octaves above the highest note audible to human ears. Each chirp is an ultrasonic tone lasting a few milliseconds (Fig. 6). In many bats it drops in pitch by about one octave during this interval, so that if it were audible to us it would sound very much like the chirp of a bird. A bat makes these chirps about 10 times a second while cruising in the open; when it encounters any object in its flight path the chirp rate may go higher than 100 per second.

The precise and ingenious experiments of Griffin[13] and his students have shown that echoes returning to the ears of the bat inform it in considerable detail about the size, distance, and location of objects in its flight path. The ground below, walls, trees, twigs, and insects smaller than midges[14] can all be located by this form of sonar. The world of a flying bat must be a series of single and multiple echoes of a subtlety that we are still far from completely appreciating. Perhaps it is more discontinuous in time than the world normally revealed by our eyes. If one walks about while opening one's eyes as briefly as possible once a second the visual world becomes a series of still pictures interspersed with intervals of darkness during

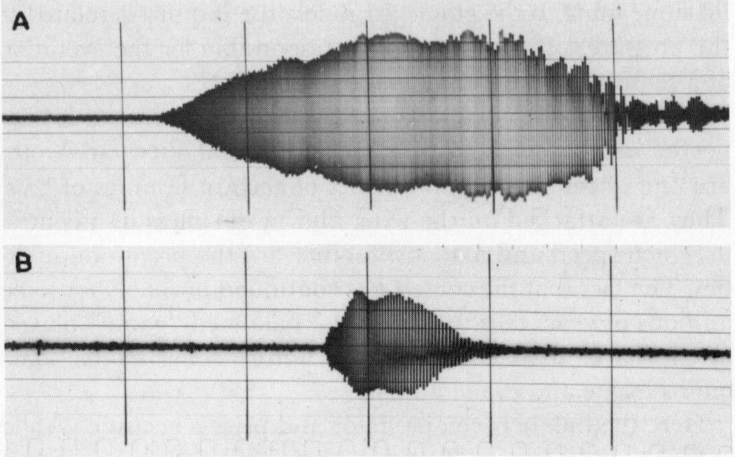

Fig. 6. Oscillograph recordings of the cries made by a bat (*Myotis lucifugus*) while locating a mealworm tossed in the air. (*A*) "Cruising" pulse; note the drop in frequency during the cry. (*B*) One of the shorter pulses containing lower frequencies made during a "buzz." The extent of the vertical deflection indicates loudness. Vertical lines mark 1-millisecond intervals. (Courtesy of J. J. G. McCue, Lincoln Laboratory, Massachusetts Institute of Technology.)

which corrective movements are made. However, in the bat's world these discontinuities would be arrayed in time because sound travels slowly compared with light. Near objects would become early echoes; distant objects would be forecast by late echoes. A single short chirp would return as a long and complex sound varying in intensity with time. The direction of this or that object must appear as differences in the arrival time of this or that component of the complex sound at the bat's right and left ears. The spatial dimensions of our visual world are converted into temporal dimensions in the acoustic world of the bat, and a flying moth becomes an intermittent and fluctuating point in time.

About 100 years ago it was suspected that moths could evade bats through a sense of hearing. The sonar system used by bats being then unknown, this was a truly inspired guess. Since that time detailed studies of the anatomy of the tympanic organ in various species and families of the Lepidoptera,[11] and observed changes in the behavior of moths in the presence of man-made ultrasound,[54,59] have confirmed the suspicion that members of certain families of moths can hear the chirps made by echolocating bats.

Assuming that the moth acts upon the information received, the advantages in survival of a bat detector are obvious. The course of its evolution is less clear. Several families of moths possess tympanic organs, including the largest families of common medium-sized moths, the Noctuidae and Geometridae. Unfortunately, not enough is known about the phylogenetic relations of these groups, although it seems probable that tympanic organs have evolved more than once. In many insects sensilla similar to those connected to the moth's eardrum are found to extend internally between jointed or elastic portions of their exoskeletons. These act as proprioceptors or "strain gauges" to measure and report on the relative movements of these parts. Reduction of the cuticle over a sensillum to the extreme thinness of the tympanic membrane might have converted it into an auditory organ.

The Tympanic Organ. The ear of noctuid moths is found on the thorax near the "waist" where thorax and abdomen join (Fig. 7). A thin eardrum or tympanic membrane is directed obliquely backward and outward into a cleft formed by flaps of cuticle, and normally covered by a thin layer of fine scales. Viewed from outside, the tympanic membrane often shows interference colors, indicating its extreme thinness.

Dissection under a microscope shows that the tympanic membrane forms the outer wall of an air-filled cavity, the tympanic cavity, that is an expanded portion of the moth's respiratory system (Fig. 8). A fine tissue strand, the acoustic sensillum, is suspended across this cavity from one point of attachment near the center of the tympanic membrane to

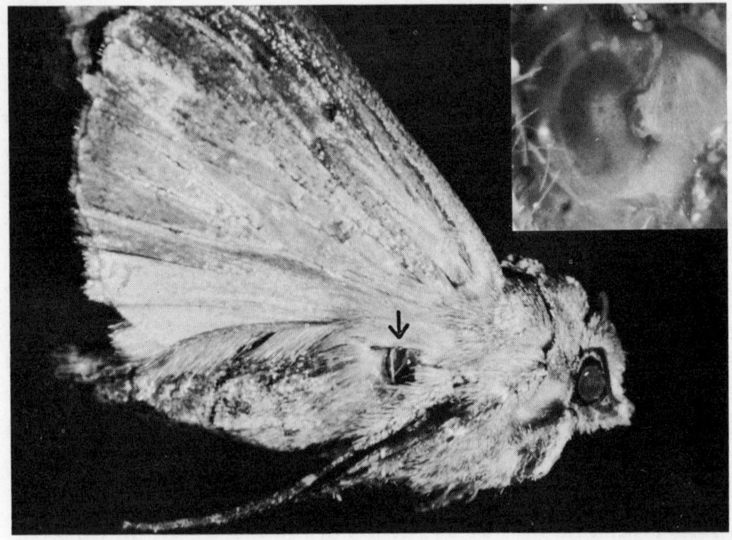

Fig. 7. External opening of the right ear in the moth *Agrotis ypsilon*.[50] The tympanic membrane faces obliquely backward and outward into the cavity below the arrow. The body of the moth is about ¾ inch in length. (*Inset*) Close-up of outer surface of the tympanic membrane as viewed through the external opening. The acoustic sensillum is visible through the transparent membrane as a white strand extending obliquely downward to its attachment near the dark spot in center.

another on a nearby skeletal support.[11,47] It is supported near its midpoint by a minute ligament attached to another part of the skeleton. The sensillum contains the pair of acoustic sense cells. Each acoustic sense cell (*A* cell) bears a fine distal process ending in a minute refractile structure, the scolops, that extends farther toward the tympanic membrane. From the central end of each *A* cell an axon passes within the sensillum toward the skeletal support, and the pair of axons continue in the tympanic nerve to the thoracic ganglia of the moth. As they pass the skeletal support the *A* axons lie close to a large pear-shaped cell (*B* cell) which may extend numerous fine fingerlike extensions into the surrounding membranes. The *B* cell gives rise to a somewhat larger axon that continues parallel with the *A* axons in the tympanic nerve to the central nervous system.

The Nerve Response. A noctuid moth, perhaps one of the common army worms whose larvae do so much crop damage, or better still a larger red underwing, is captured as it flies to a nearby light. Under temporary anesthesia it is decapitated and firmly restrained on the stage of a dissecting microscope with small strips of plasticine in such a position that the tympanic openings have an unrestricted sound field. The scales on the thorax are removed with a small paint brush, and the dorsal part of the thorax, including one of the main

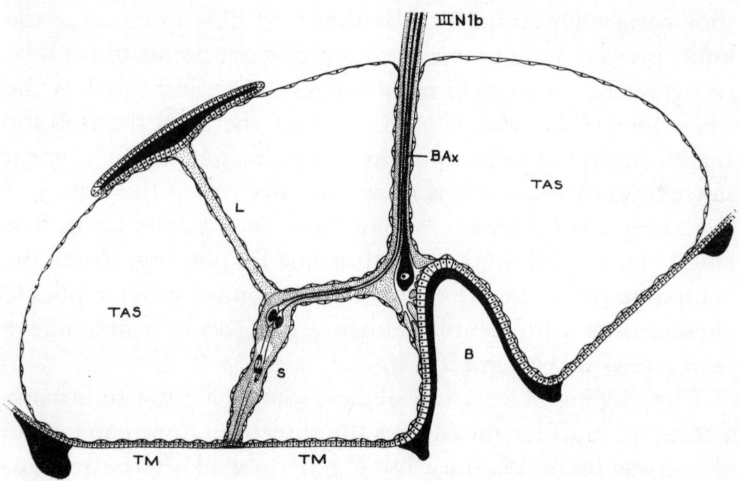

Fig. 8. Diagram of the tympanic organ of a noctuid moth.[61] The sensillum *S* contains the pair of acoustic receptors or *A* cells. It is attached to the tympanic membrane *TM* and suspended in the tympanic air sac *TAS* by a ligament *L* and by its nerve passing to the skeletal support *B*. Here the *A* nerve fibers are joined by the *B* cell, and run together with the *B* nerve fiber *BAx* to form the tympanic nerve *IIIN1b*. Activity was recorded from the tympanic nerve after it had left the tympanic air sac.

sets (horizontal) of flight muscles, is dissected away. The tympanic nerves run forward on either side of the cavity thus revealed, passing from the tympanic organs at the back of the thorax to the large pterothoracic ganglion that supplies all the organs of the thorax.

There are several nerves in this region, all small and transparent. However, the task of hooking a tympanic nerve on an electrode is not as hard as it might seem. One electrode is a silver wire inserted anywhere in the tissues of the moth. The other is a silver wire tapered to a fine point that is bent into a minute hook. This active electrode is manipulated mechanically. Both electrodes are connected to an amplifier and cathode-ray oscilloscope, also to a loud-speaker. Since spike potentials cause minute brief current pulses at the electrode, they can, when amplified, be made audible as clicks in the loud-speaker. When a nerve contains active axons this can be recognized by a series of regular clicks when it is lifted by the silver hook. The hook is used to "fish" the body fluids in the region suspected to contain the tympanic nerve. Each time a nerve is encountered it is raised slightly out of the fluid and the experimenter makes "sshing" sounds or jingles his keys or small change, all sources of ultrasonic frequencies. When the tympanic nerve has been hooked, the loud-speaker replies to these sounds with a rapid sequence of clicks that may merge into a low-pitched musical tone.

The jingling of keys and sibilant sounds are not sufficiently precise stimuli for measuring the discriminatory capacity of the *A* receptors. We must use a generator of alternating current that can be controlled in frequency and voltage, and a loud-speaker or transducer to convert the current into sounds that can be varied in pitch and loudness. Furthermore, the moth must be shielded from incidental sounds—somewhat of a problem since it can hear sounds inaudible to us.

In the absence of sound the pattern of spikes transmitted by the axons of the tympanic sense cells appears to be somewhat random. The most regular feature of the record is the larger spike, which belongs to the *B* cell. It repeats with almost clocklike regularity, commonly between 10 and 20 times a second. In some cases the *B* cell is inactive; rarely it may discharge 300 impulses per second. The smaller spikes appearing at irregular intervals belong to the *A* axons. This level of spontaneous impulses entering the moth's nervous system must signify "no bat."

Intensity. When the silence is broken by continuous pure ultrasonic tones of various intensities the typical response of the *A* cells is that shown in Fig. 9. At the onset of a very faint tone (*A*) there is a small burst of *A* spikes which immediately tails off into an irregular sequence. At a higher intensity (*B*) the initial frequency of *A* spikes is greater, and a regular discharge continues during the tone, though with declining frequency. At a still greater sound intensity (*C*) the *A*-spike frequency is increased and better maintained, but it still declines as the tone continues. Also, occasional spikes appear to have double peaks. At the highest sound intensity used in this experiment (*D*) the nerve response becomes quite complex, with many spikes, double peaks, and spikes that appear to have double the normal height. In all the records the much larger spike potential of the *B* cell appears infrequently but at regular intervals, and is completely unaffected by the ultrasonic stimulation.

Inspection of these records shows that sound intensity is coded by the *A* cells as number of spikes per second. A count of the number of spikes (not including the *B* spikes) in the first 0.1 second of each tympanic nerve response gives in (*A*) 9 spikes, in (*B*) 39 spikes, in (*C*) 62 spikes, and in (*D*) about 78 spikes. The coding of sound intensity in spike fre-

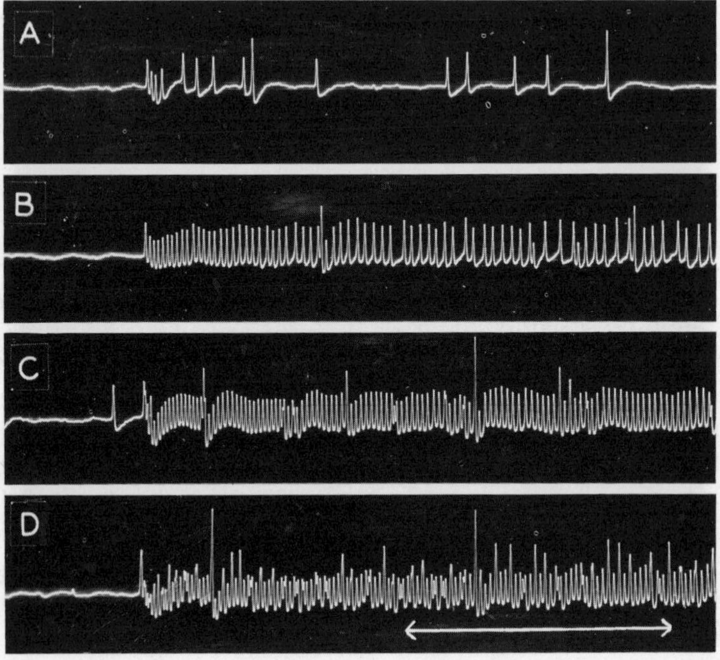

Fig. 9. Tympanic-nerve response in the moth *Prodenia eridania* to a pure tone of 40 kc/s.[38] The occasional large spikes originate in the *B* cell. (*A*) Response to a sound intensity close to the threshold of the most sensitive *A* cell. The sound intensity in (*B*) is 7 db, in (*C*) 15 db, and in (*D*) 23 db above that in (*A*). Time line in (*D*), 100 msec.

quency is accomplished in two ways. In records (*A*) and (*B*) the evenness of the spike sequence (allowing for irregularities in the base line) and the absense of double peaks suggests that all the *A* spikes are being transmitted in a single *A* fiber. In record (*C*), and to a greater extent in (*D*), double peaks and spikes of extra height become more frequent. The all-or-none nature of the spike in a single fiber and the refractory period that must accompany it (Chapter 3) would make it impossible for a single fiber to produce a double-peaked action potential and to produce action potentials of abruptly varying heights. The only possible conclusion is that when the sound reaches a certain intensity, actually about ten times that causing a response in one *A* cell, the other *A* cell begins to respond in like manner. Since the electrode on the nerve detects all three axon transmissions (two *A* and one *B*) simultaneously, they are added together in the electrical record. However, it must be kept in mind that each is being transmitted along its own axon as a separate sequence to the central nervous system, and that this superposition is an artifact of the method.

Another point of interest in Fig. 9 is the decrease in spike frequency that takes place as the sound continues. If intensity is coded as number of spikes per second, first in one *A* cell and then in both working simultaneously, this decrease in frequency with the passage of time must mean that the sound is represented to the moth as becoming progressively fainter, even though it has remained physically unchanged. This progressive loss in sensitivity is known as sensory adaptation, and is actually widespread and familiar in everyday experience. If adaptation did not occur in most receptors registering changes in the outer world, the impact of our surroundings would be often shocking and unbearable. The brilliance of a lighted room entered after dark would remain blinding, and the contact of our clothing with our skin would irritate the day through. The speed with which receptors adapt varies greatly—the moth's *A* cells adapt relatively rapidly—and some adapt slowly if at all. The latter types are most concerned with reporting the positions of our limbs (Chapter 8) and our position with respect to gravity. If these did not give continuous and faithful reports on constant conditions we would be unable to remain motionless or upright for any length of time.

From this it can be inferred that a moth would be unable to discriminate a continuous tone from a short pulse, particularly at low sound intensities, although it could discriminate differences in sound intensity. Since the tympanic organ presumably serves as a bat detector, and bat chirps are short sound pulses, it is necessary to examine the *A* responses to artificial sounds similar to bat cries.[42]

In the experiment shown in Fig. 10 a single short ultrasonic pulse was generated artificially at regular intervals. It was similar to a bat chirp except that it lacked the frequency modulation of the natural sound. A microphone was placed near the tympanic organ of the moth. In each frame the upper line traces the electrical response of the microphone

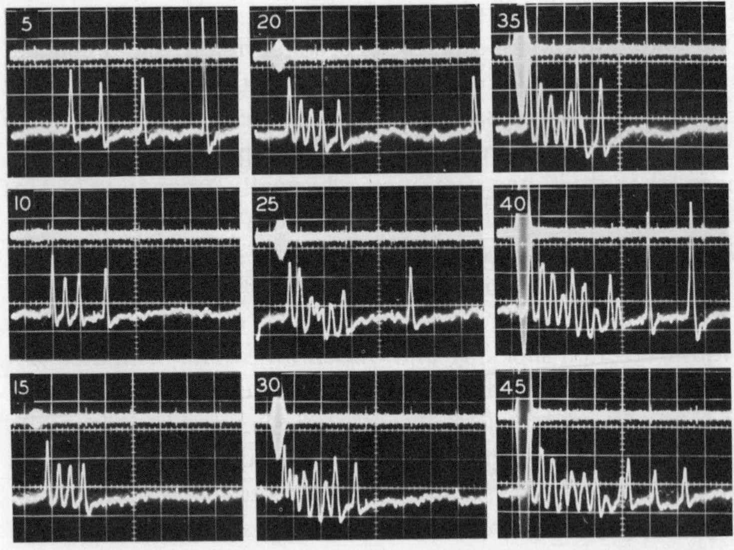

Fig. 10. Artificial sound pulses of 70 kc/s recorded by a microphone (*upper traces*) and by the tympanic organ of the moth *Noctua c-nigrum* (*lower traces*).[50] The number on each recording indicates the intensity in decibels of the sound above a reference level close to the threshold of the more sensitive *A* receptor. The less sensitive *A* receptor first responds in the 25-db recording. Large spikes in some of the recordings belong to the *B* cell. Vertical lines are 4 msec apart.

and the lower the neural response of the tympanic organ to a given pulse. The intensity of the sound pulse was adjusted until it just failed to produce a response in the *A* cells (0). It was then increased in measured steps of 5 decibels, and the microphone and nerve response were recorded at each step. As before, the increase in sound intensity is accompanied by greater crowding of the *A* spikes, and double peaks—evidence of activity in the second, less sensitive, *A* cell—appear when the sound is about 25 decibels above the threshold of the first. Closer inspection of the records shows two additional ways in which the *A* response changes with increasing sound intensity. First, the sound lasts only 3 milliseconds, as shown by the microphone record. Yet at the higher intensities the spike discharge continues for several milliseconds after the sound has ceased. It is as if the more intense sounds cause in the sense cells some overaccumulation that continues to generate impulses after the sound has ceased. The nature of this overaccumulation is unknown, but the length of the after-discharge is clearly determined by the intensity of the sound. However, it is worth noting that a sound of lower intensity lasting for a longer time produces the same response, and that the moth must be unable to distinguish between the two.

The second additional intensity-coding device revealed in Fig. 10 is less obvious but quite important. The sound was timed to occur always at the same instant in relation to the vertical time lines on each frame. It can be seen from the position of the first spike potential in each frame that the response begins earlier as the intensity is increased. Since the velocity of the impulses cannot have been affected by the stimulus strength, it must be concluded that the response time in the sense cells, that is, the interval between arrival of the stimulus and generation of the impulse, must decrease with increasing stimulus strength. This change in response time, which amounts to between 1 and 2 milliseconds, could be of no use to the moth unless it had an additional measure of the arrival of the stimulus. This could be provided by the tympanic organ on the other side of the body, and may have significance in location of the direction of sound sources.

In Fig. 11 results of other experiments on some of these means for measuring loudness are summarized. The number

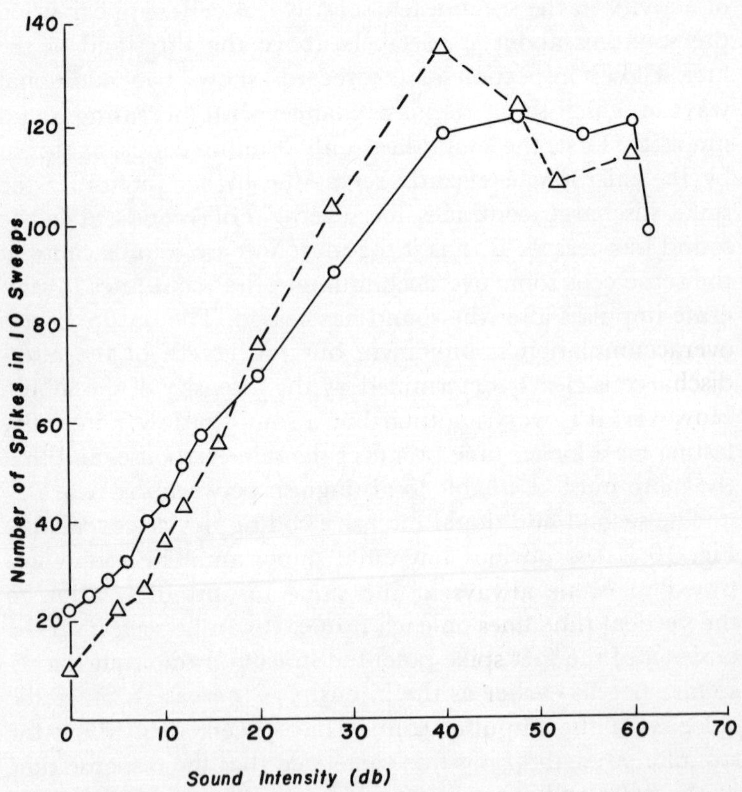

Fig. 11. The relation between the intensity of a 5-msec sound pulse of 50 kc/s and the number of *A* spikes in the response, for two specimens of the moth *Prodenia eridania*. Each point is the sum of 10 separate measurements. Sound intensity is in db above threshold intensity.

of spikes elicited by ten sound pulses at each of a number of intensities was plotted against the sound intensity relative to that at the threshold of the most sensitive *A* cell. It can be seen that the number of spikes in the *A* response bears a constant relation to sound intensity over a range of roughly 40 decibels or about 100-fold. At higher intensities than this the *A* cells saturate, that is, they are unable to produce further increases in the after-discharge.

The main points in this analysis of the *A*-cell response to sounds of different intensities are:

(1) Some *A*-cell activity continues in the absence of ultrasonic stimulation;

(2) The *A*-cells adapt rapidly to a continuous pure tone;

(3) The intensity of an ultrasonic pulse is coded in the *A*-axon discharge as (i) the number of *A* spikes per second, (ii) the activity in one or in both *A* cells, (iii) the duration of the after-discharge, (iv) the response time;

(4) Sound intensity is coded in these aspects of the axon discharge only within a certain range of sound intensities, beyond which the *A* receptors saturate.

The significance of these findings lies not in any special or unique properties of the *A* receptor cells. On the contrary, it is important to note that these properties are classical, that is, similar to those reported many times previously in many animals by many observers, but heretofore only in single units isolated for experimentation from a complex sense organ containing many thousands of receptors. They become significant only when it is remembered that they encompass the whole sense organ, not merely a small part of it, and thus define the total sensory input being communicated to the effector mechanisms for the evasion of bats. Before turning to this side of the story it is necessary to define more closely the function of the tympanic organ by mentioning a few more experiments, mostly negative in outcome.

Pitch. The pitch of a pure tone is determined by the number of times per second the air is compressed and rarefied by the source. The adult human ear normally detects tones ranging in frequency from about 40 to 20,000 cycles per second or 20 kilocycles per second (kc/s). Complex tones—and this includes most natural sounds such as bat cries—contain a fundamental frequency overlaid by various overtones and harmonics. The echolocating chirps of the common insectivorous bats contain fundamental frequencies ranging from 80–100 kc/s down to 20–25 kc/s, and thus lie outside the normal range of human hearing, the upper limit of which is about 20 kc/s. Faint clicks can sometimes be heard as a bat swoops close by, and audible cries are often made by a bat held in the hand.

Stimulation of the moth's tympanic organ with pure tones of different pitch shows that it can hear sounds from about 3 kc/s, well within the human hearing range, to over 150 kc/s, the practical limit of the equipment available to us for producing and measuring ultrasound.[47] It responds well to whistling, singing, and the higher vocal sounds, as well as to a wide variety of rustles and clicks. If actual sound energies are measured, however, its greatest sensitivity lies near the middle of this range, that is, 50–70 kc/s.

There is no evidence that the *A*-cell mechanism can discriminate between tones of different pitch. Crude pitch discrimination would be theoretically possible in such a simple mechanism if the *A* cells differed from each other in the pitch to which they were maximally sensitive. There is no evidence that this is so, and it must be concluded that noctuid moths are tone deaf.

The B Cell. The *B* cell was searched for, and discovered lying in the membranes near the cuticular support of the acoustic sensillum, only because the regular beat of its large spike makes such a striking contribution to the activity of the tympanic nerve.[61] In spite of the fact that the *A* axons pass very close to the *B* cell, there seems to be no interaction between *A* and *B* impulses. The *B* cell continues its regular discharge of between 5 and 15 impulses per second during the most intense acoustic stimulation (Fig. 12*A*). In simultaneous records from right and left tympanic nerves each *B* cell discharges spikes at its own rhythm (Fig. 15).

Decapitation, disconnection of the tympanic nerve from the central nervous system, and destruction of the tympanic membrane and the *A* cells within the sensillum all fail to alter the rhythm of *B* spikes. Changes in the pressure, temperature, and composition of the gas within the tympanic air sac

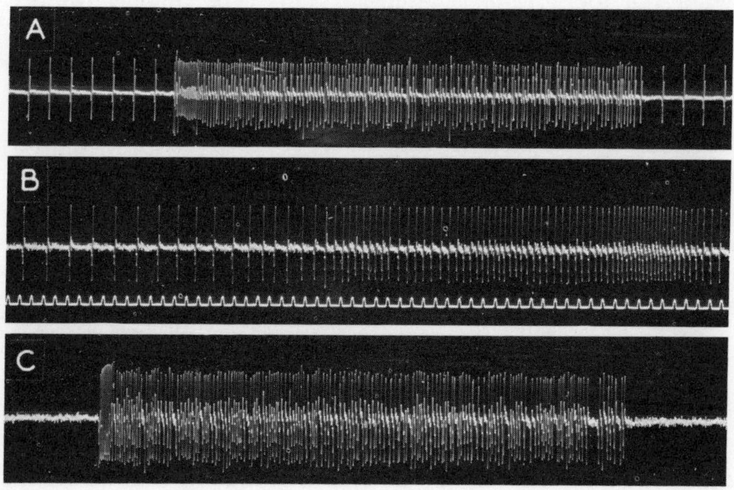

Fig. 12. Activity of *A* and *B* fibers in the tympanic nerve of *Catocala* sp. (*A*) A pure tone of 26 kc/s causes a vigorous response in the *A* receptors without altering the rhythmic discharge of the *B* cell. The *B* spikes can be seen extending above the *A*-cell response. (*B*) No acoustic stimulation; the *B*-cell rhythm is being altered by manipulating the membrane near the *B*-cell body. (*C*) Acoustic response as in (*A*), but after destruction of the *B* cell. Time signals in (*B*), 10 per second. (Partly after Treat and Roeder.[60])

affect the *B* activity only to the extent that these measures would affect any nerve cell. Rotation or acceleration of the whole moth likewise give negative results.

Mechanical manipulation of the tympanic organ and its surroundings is the only known cause of consistent changes in the *B*-spike frequency.[61] Gentle traction on the membranes lining the air sac, or pressure on the surrounding skeletal members, will increase the *B*-spike frequency up to a maximum of 200 to 300 impulses per second (Fig. 12*B*). Unlike the *A* response to acoustic stimulation, the *B* response to mechanical stimulation adapts slowly, if at all. The increased frequency of discharge persists as long as traction on the membrane is maintained, and drops to a slower rhythm only when the membrane is released. This type of performance hints that the *B* cell is a proprioceptor, measuring internal displacements and stresses in the region of the ear that are likely to accompany movements of the wings during flight. Perhaps it is a remnant of the proprioceptive organ from which the tympanic organ is thought to have evolved (page 37), but it is obviously still functional and active. It is not known whether its presence in the tympanic organ and its proximity to the *A* axons is incidental, or whether it plays some specific role in connection with hearing.

The Sensory Event. Having proceeded thus far in a physiological analysis of the operation of the moth's auditory organ, it is tempting to ask questions about the manner in which acoustic energy is transduced into nerve impulses by the *A* cells. Inquiry into these decision-making or discriminating mechanisms would carry us to a finer level of analysis and further from our objective of fitting the neural mechanisms into a behavioral context. However, the answers are lacking, so it might be worth while framing a few of the questions in the hope of exciting someone's curiosity.

How do 100,000 infinitesimal vibrations per second in the

tympanic membrane bring about the totally different, much greater, and much slower energy flux of ions across the *A*-cell membrane? The process is vastly more mysterious than the conversion of acoustic energy into electric energy by a microphone. The 10-micron *A* cell is at the same time microphone, rectifier, amplifier, and power-conversion unit combined. How is it that the *A* cell, or for that matter any sense cell, is specialized or perhaps shielded so that it responds only to one type of external change—light, sound, chemicals, or displacement—and not to others? How can the ultrastructure of the scolops help in understanding these matters?

These questions seem to carry us away from the behavior of the whole animal, but they directly concern the decision-making or discriminating properties of nerve cells discussed in Chapter 8, and must eventually be answered. For the moment we must be content with asking how the decisions made by the *A* cells and communicated in *A* spikes to the moth's nervous system play a part in its behavior.

5. *Moths and Bats*

These experiments with artificial sounds have introduced the elements of vocabulary and grammar of the neural language. Fortunately the tympanic organ communicates with the moth's central nervous system only in the simplest form of this language, so that even after this elementary instruction it is possible to interpret some biologically significant messages. These are the chirps made by bats while going about their natural occasions.

The first record of a tympanic nerve response to the chirps of a flying bat was obtained in the laboratory, and almost by accident.[47] Experiments with artificial sounds were in progress during the month of January, a time of year when New England bats are deep in hibernation. A student making a week-end exploration of a New Hampshire cave found a hibernating bat and brought it back to the laboratory, where it was placed in a refrigerator and almost forgotten for several weeks. When removed and held in the hand of an experimenter near a tympanic-nerve preparation and microphone, the bat recovered sufficiently to deliver a few angry and audible shrieks (Fig. 13*A*), and an energetic bite. This naturally brought about its release, whereupon it flew "silently" round the laboratory close to the ceiling. Throughout the flight the tympanic nerve delivered a rapid series of short bursts of *A* spikes. When the bat flew sufficiently close to the experimental table the microphone joined in with its electronic version of the ultrasonic chirps (Fig. 13*B*).

This impromptu experiment showed not only that the tympanic organ responds as expected, but also that it is highly sensitive to bat cries. One or both of the *A* fibers continued to respond at times when the bat was too distant for its cries to register in the microphone. The moth could hear the bat at all points within the laboratory, and we were most eager to go beyond its walls and into the field.

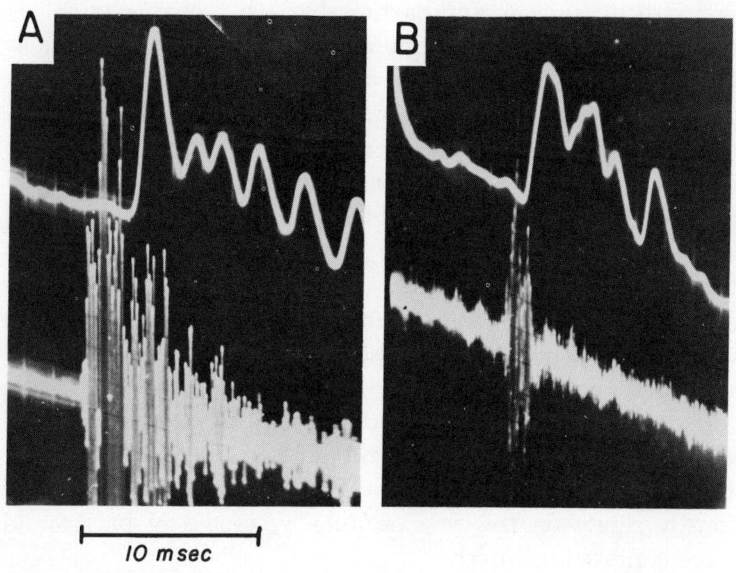

Fig. 13. Tympanic-nerve responses of the moth *Prodenia eridania* to cries made by the bat *Eptesicus f. fuscus.*[47] The cries were also simultaneously registered by a microphone (*lower trace*). (*A*) Response to an alarm cry audible also to our ears; the bat was held about 2 feet from moth preparation and microphone. (*B*) Response to a short ultrasonic cry made by the bat while in flight. Note that both *A* cells respond in both records.

This turned out to be somewhat more than the carefree jaunt it suggests.[49,50] About 300 pounds of electronic gear was hauled up a grassy hillside in the Berkshires of Massachusetts, and reassembled in a spot where bats were known to feed. At dusk a moth was captured at a nearby light and mounted so that one tympanic organ had an unrestricted sound field. Its tympanic nerve was hooked on an electrode and the *A* and *B* fiber activity was followed continuously on an oscilloscope and loud-speaker. The spikes were also recorded on magnetic tape.

A part of one of the records made on this first evening is shown in Fig. 14. It was dark before all was ready, and the first indication of an approaching bat was a change of the *A*-spike patterns from an irregular sequence to regular bursts at 10 per second. When heard via the loud-speaker the bursts appeared to rise in pitch and change in quality, then to go through the reverse sequence and fade away. This was interpreted as the approach and departure of one bat. The approach can be seen in Fig. 14*A*. At maximum range the bat chirp evoked a single *A* spike (following a large *B* spike). As the bat approached, probably on a wavering course, successive chirps evoked 4, 3, 8, 9, and 11 spikes. Both *A* cells were active (double peaks) in the last three bursts. In each succeeding burst the spikes were packed closer together, giving the burst a higher tone when heard through the loud-speaker. It did not take us long to learn to read from these sounds something of the movements of the bats. Luckily they flew over mostly singly on this first evening. A crowd of bats would have confused us utterly.

Other spike patterns soon became recognizable. In Fig. 14*B* each burst is very long, and some appear to be double. These were undoubtedly evoked by a bat flying very close to the preparation, and the only explanation of the second group of spikes in each burst is that the moth ear detected not

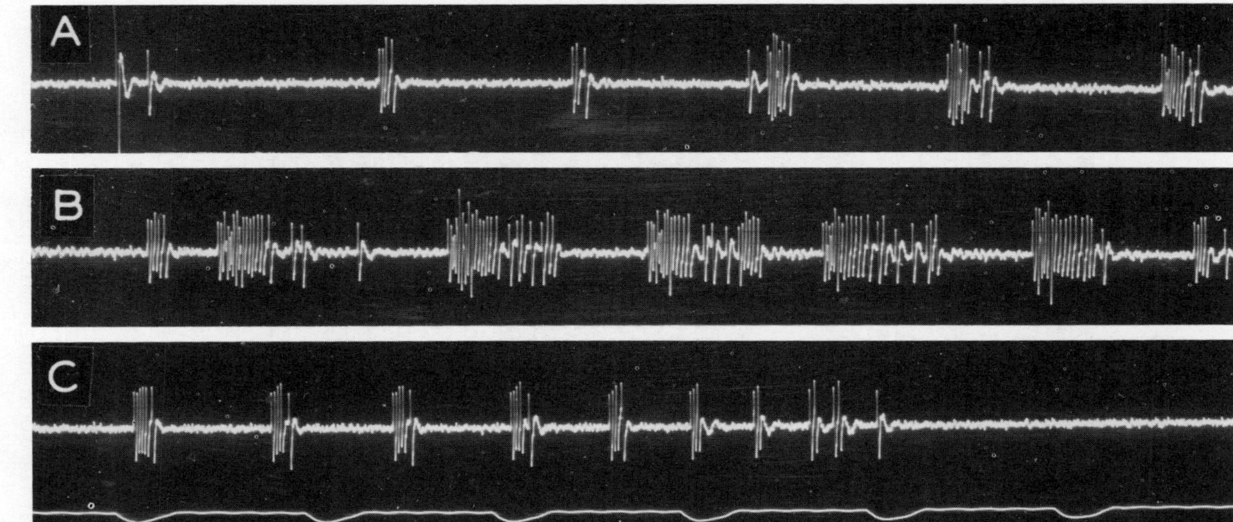

Fig. 14. Tympanic-nerve responses of *Noctua c-nigrum* to the cries of bats flying in the field.[50] (*A*) The approach of a cruising bat emitting pulses about 10 times a second. (*B*) Responses to a bat flying near the preparation. The moth ear detects each cry as well as its echo (*second impulse volley*) from a nearby wall. (*C*) Tympanic response to a "buzz." Time marks in (*C*), 10 per second.

only the bat's chirp, but also its echo from a nearby wall. This was confirmed by comparing the distance ($\times$ 2) of the preparation from the wall with the distance sound could have traveled in the time interval between the two bursts of the double records.

At intervals the cruising rhythm of about 10 bursts per second changed to a rising crescendo. Records of this change showed a decline both in the interburst period and in the number of spikes in each burst (Fig. 14*C*), and is sometimes terminated in a continuous train of *A* spikes (Fig. 15*C*) followed by an interval of inactivity. This pattern was apparently caused by the "buzz," or greatly increased frequency of chirps emitted by a bat when it detects an echo from a nearby object such as a flying insect. The buzz continues as the bat tracks and closes with its prey, so that this pattern of *A* spikes is probably the last ever registered by the tympanic organs of some flying moths.

The high excitement of listening for the first time to these night sounds through a moth's ear was tempered by the thought that we had no independent evidence that they were being caused by bats. They were inaudible to us, and we had with us no ultrasonic microphone to provide a separate record. A floodlight was rigged so that we were able to observe bats flying within 20 feet of the preparation. It then became clear that the range of the moth ear was much greater than that of the light, so that the appearance of a bat in the lighted area could often be predicted by listening to the rising pitch of successive *A* bursts. It was difficult to establish the range of this biological bat detector, since this depends upon the species of moth and bat as well as the relative angle of their flight paths. In another experiment[49] a moth preparation was set up at dusk about 200 yards distant from an old barn where bats roosted. It was known that at this hour the bats usually left the roost singly and flew on a straight path directly over the site chosen for the preparation to other feeding grounds. An observer, wearing headphones connected by a long cord with the amplifier, walked "upstream" toward the barn while listening for the first signs of regular *A* bursts and watching the bats passing overhead. The range of the moth ear appeared to be 100 feet or more. A later experiment, planned to measure the range more accurately, is described in Chapter 6.

Binaural Nerve Responses. All this information came from one ear of a moth. What could be learned by recording from both ears simultaneously? This project had to wait until the following summer, for it was necessary to learn how to insert and manipulate two hooked electrodes within the small space of a moth's thorax, and to duplicate most of the amplifying and recording equipment. The activity in right and left tympanic nerves was recorded on stereo magnetic tape, and subsequently photographed by replaying the tape into a dual-beam oscilloscope.

A binaural nerve response to a flying bat is shown in Fig. 15. Two active electrodes under the tympanic nerves were pitted against a common "indifferent" electrode placed in the abdomen of the moth. The large, slow, diphasic waves appearing on the records from both nerves are the moth's electrocardiogram, and have no connection with the tympanic-nerve responses. Spikes in the *B* fibers, distinguished by their greater height, recur regularly but quite independently in the right and left activity traces.

The approach of a bat is signaled first by two *A* spikes (immediately following the first *B* spike) in the upper trace of the first recording (Fig. 15*A*). The other ear (lower trace) does not detect the bat until the next chirp, when two spikes indicate lower intensity in this ear compared with four spikes registered by the other. This difference is maintained with the third response, but by the fourth (all of which is shown) there

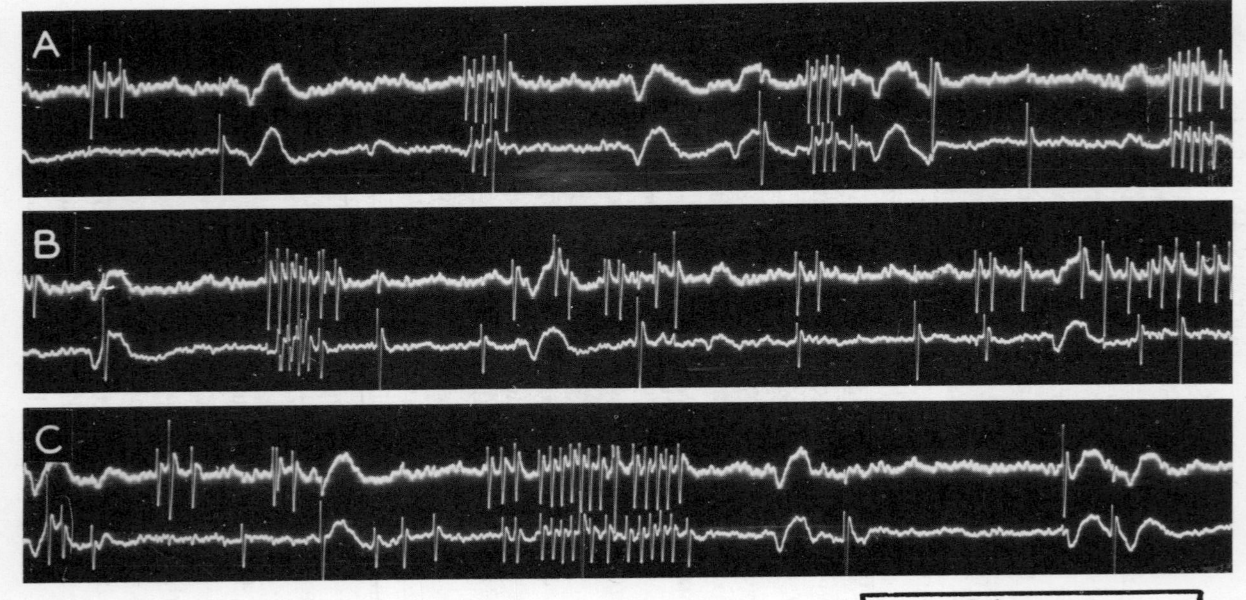

Fig. 15. Binaural tympanic-nerve responses of *Feltia sp.* to the cries of red bats flying in the field.[50] The slow waves on both channels are the electrocardiogram of the moth. *B* impulses (large spikes) appear regularly but without synchrony in both traces. (*A*) An approaching bat. Differential tympanic response (latency, number of spikes) between right and left is marked at first, but has practically disappeared in the final response. (*B*) A "buzz" registered mainly by one ear. (*C*) A "buzz" registered a few seconds later by both ears.

is little difference between upper and lower traces. This suggests a bat approaching from one side of the moth until it is directly overhead.

The binaural nerve response contains another differential determined by intensity that was not present in the monaural recording. This is the difference in response time between the right and left bursts. The first spike in the second burst of the upper trace precedes the first spike in the corresponding burst of the lower trace by a millisecond or so. In the following (third) burst the difference in time of appearance of upper and lower first spikes is less, while in the fourth and last burst of this record it has practically disappeared, the initial spikes being almost simultaneous. In the experiments with artificial sound pulses (page 45) it was shown that the absolute response time of the *A* cells decreases with increasing sound intensity. Also, it was pointed out that this criterion of intensity would not be available to a moth having only one ear. However, in an intact insect the sign (left or right) and duration of the differential response, together with the differential spike frequency and number, could tell the moth whether its adversary was approaching from the right or left.

With this in mind the other records of Fig. 15 suggest various movements of the bat, which seem to have taken place mostly toward that side of the moth represented by the upper trace. A buzz occurs on this side toward the end of record (*B*), followed in (*C*) by another buzz emitted when the bat was almost overhead. This is followed by a silence that may indicate the end of a successful pursuit by the bat.

It is interesting to listen through stereo headphones to the taped responses of right and left tympanic nerves to a moving bat. The human ear interprets these spike differentials as giving direction to the source, and one can almost imagine oneself inside the nervous system of the moth as the source of clicks appears to move from one side to the other. This illusion of direction is not continuous, and much of the time the source of sound seems to be in the center of one's head. The explanation is that the spike differential is greatest at low chirp intensities, becoming less and disappearing above a certain loudness. This is shown in Fig. 15*A*, and also by the measurements of Fig. 11, from which it can be seen that sounds louder than 40 decibels above the threshold of the most sensitive *A* cell can elicit no further increases in the *A* response. This saturation of the acoustic response above certain sound intensities indicates that a moth would be better able to determine the bearing of a bat when the latter was near its maximum range of hearing.

This differential response would be possible only if the ears of a moth were somewhat directional, responding better to sounds on one side of the body than on the other. This directionality was measured directly for one tympanic organ in the relatively echo-free situation of an open meadow.[49] A loud-speaker producing clicks of constant intensity was moved along radii to the moth at 45° intervals. On each radius that distance was established at which the click produced a standard spike response arbitrarily chosen to be slightly above the threshold of the preparation. Eight radii, covering 360° around the moth, were mapped in this manner. A polar graph of the distances showed that, while there was little difference fore and aft, a click on the near side was heard at about twice the distance of a symmetrically placed click on the far side. More detailed measurements are given in Chapter 6.

This information extracted from the tympanic-nerve responses makes possible a crude prediction of the moth's behavior upon detecting an echolocating bat. If it is assumed that a bat is first detected at a distance of 100 feet and then approaches on a straight path at right angles to the moth's course while making chirps of constant intensity, the differential tympanic response would decline from a maximum at about 100 feet to zero at 15 to 20 feet. Within this range the moth would have sufficient information to enable it to turn away from the bat. At a range of less than 15 to 20 feet the neural information reaching the moth's central nervous system would make possible only nondirectional responses.

Avoidance Behavior of Moths. This is as far as we can go at present in assessing the acoustic information coded and transmitted to the moth's central nervous system by the *A* cells. The next step would be to find out what happens to the *A* impulses after they have entered the pterothoracic ganglion —the main coordinating center of the thorax—and to trace their effects on the neurons that must eventually connect with the muscle groups used in avoidance behavior. Progress has been made in working out the neural circuitry concerned in moth avoidance behavior (Chapter 11), but until some of the principles of nerve cell interaction have been examined it is better to regard a moth's central nervous system as a "black box" and turn to a study of its output.

The prediction reached from examining the *A*-cell responses to bat cries is of the sort most of us first made in our childhood: "what I would do if I were a moth." In order to reach it the sensory information coded in the *A* spikes was processed in my central nervous system, not that of the moth. The latter may be able to extract either more or less information from its tympanic nerves than I have been able to do. Furthermore, there is no assurance that the tympanic nerves are a moth's sole source of information on the whereabouts of bats, or that moths do not react to sounds other than those made by their predators. It is time to climb back out of this neurophysiological blind alley and observe the behavior of free-flying moths.

During the past fifty years, and probably much earlier, many people[11] have noted that the flight movements of certain moths become rapid and erratic in the presence of high-pitched sounds. The squeak of a glass stopper, the jingle of keys or coins, the high notes of a violin or flute, and a variety of rustling and hissing sounds are effective. On a summer evening anyone can observe this behavior by suddenly jingling a bunch of keys while watching the moths flying near a street light, or fluttering on the screen covering a lighted window. Some of the moths will certainly respond.

It is easy to show that some moths respond to high-pitched sounds, but somewhat harder to describe just what they do. Some fold their wings and fall to the ground, while the flight of others becomes faster and more erratic; some fluttering individuals become motionless, while inactive moths may take flight. Several workers[54,59] have examined these reactions more closely by restraining or tethering moths and registering their flight movements, and by using more precise sources of sound. These experiments have established that reactions to ultrasound occur only in those species known to have tympanic organs, and then only when one or both tympanic organs are intact; also that the most effective sound frequencies correspond approximately to the frequencies in the echolocating cries of bats.

Similar reactions can be readily observed in moths being chased by bats. It is only necessary to find a place where bats regularly feed, usually any grassy area where they repeatedly swoop and circle after dark. A minimum amount of illumination, perhaps a 100-watt bulb with reflector, and a fair amount of patience and mosquito repellent are all that is required.

The "unaltered" flight patterns of moths differ greatly in different species. Some dash rapidly across the lighted area, while others fly straight into the light. The easiest to observe are those that move across the area with a slow or wavering, almost hovering, flight. As a bat comes "silently" out of the darkness the flight pattern of the moth suddenly changes to any one of a number of maneuvers—dives, rolls, repeated tight turns, or rapid flight just above the ground. The bat may make a single pass or turn at once to make another, or it may attempt to follow the moth through its gyrations. It is a dizzy dogfight. Extrapolation of a string of acoustic dots in time is pitted against unpredictability; power and speed against maneuverability. The details may be difficult to discern, but the outcome is seen either as a bat and a moth going their separate ways, or a departing bat and moth wings fluttering slowly to the ground.

We made an attempt to find out the extent to which the odds in this contest are influenced by the avoidance tactics of the moth.[48] We observed 402 encounters between bats and moths and scored for the presence or absence of a sudden change in the flight pattern of the moth as the bat approached and for the outcome—capture or escape of the moth. Analysis of the pooled data showed that for every 100 reacting moths that survived attack only 60 nonreacting moths survived, meaning that the selective advantage of evasive action was 40 percent.

These figures focus upon only one instant in the life of a moth, although certainly an important one. At other times it is possible that the possession of tympanic organs and evasive mechanisms weigh differently, even negatively, in survival, so this measure does not describe the over-all survival advantage of possessing tympanic organs. Nevertheless, it seems adequate to account for their evolution.

It is not easy to tell at what instant a cruising bat first detects a medium-sized moth and turns to the attack. The actions of both contestants are so rapid and unpredictable to the observer that is has not yet been possible to capture the whole contest on slow-motion movie film together with a sound track of the bat's chirps, although this has been done with inert targets tossed mechanically into the bat's path in place of the flying moth.[65] A multiflash exposure (Fig. 16) gives

Fig. 16. Multiflash photograph of a bat closing with a falling mealworm. Successive positions of the bat and its target are shown by the numbers. The bat's mouth faces the target as it drops, but its flight path follows an intercepting course. Capture occurs between *3* and *4*. (Courtesy of Frederic A. Webster.[13,65])

some idea of the maneuvers made by an attacking bat. It seems unlikely that the bat makes acoustic contact at distances greater than 10 to 15 feet. The tympanic-nerve studies showed that within this range the average bat cry is capable of saturating both ears of a moth, so that the latter can make only nondirectional responses.

Yet the tympanic organs can detect a bat cry at much greater distances (see Chapter 6). At these ranges there is a marked difference in the nerve responses of the right and left ears when the bearing of the bat is to the right or left of the moth. There would seem to be little survival advantage to the moth in making erratic turns and twists when the predator was still so distant. They could be of value only at close quarters when the small inertial moment and short turning radius of the moth is pitted against those of the more massive bat. The neurophysiological data suggest that moths might show directional responses to the cries of distant bats. This had not been reported from field observations, so a field experiment was designed to find out whether moths did in fact show two sorts of behavior to ultrasonic stimuli.[40]

The complexity of the natural situation, in which both sound source and detector are continually on the move, was reduced by replacing the bat with a stationary multidirectional transmitter of ultrasonic pulses. This was mounted on a 16-foot mast at the edge of a lawn surrounded by low vegetation. The observer was seated 25 feet away behind a floodlight that illuminated a broad area of garden and silhouetted the transmitter on its mast against the night sky. This view of the transmitter was also framed in the field of a 35-mm still camera.

The observer had at hand two switches, one controlling the ultrasonic signal and the other the camera shutter. When a moth was seen to move into the field of the camera the shutter was opened and the moth's track was recorded as a continuous line against the black background of the sky. Undulations on the line were caused by the moth's wing movements. After a stretch of flight track had been recorded the switch controlling the ultrasonic signal was depressed. This released a train of ultrasonic pulses, commonly at a rate of 30 per second, each 5-millisecond pulse having a frequency of 70 kilocycles per second. The pulses were shaped as much as possible like bat cries, but they lacked the frequency modulation of the natural sound.

Thus, the moth's flight path was recorded before and during ultrasonic stimulation. The onset of the pulse sequence is shown by an extra-bright spot on each record, while the timing of events is indicated by gaps repeated at quarter-second intervals throughout the track.

The worst defect of the method is the large amount of light needed to secure a satisfactory photographic record. This might have altered the responsiveness of moths to a signal they normally encounter only in darkness. However, visual observations under illuminations too low for photography, and with yellow and red light to which moths are much less sensitive than man, showed no substantial differences in the behavior. A second problem was the difficulty of identifying the moth species producing the tracks. Many flew away before

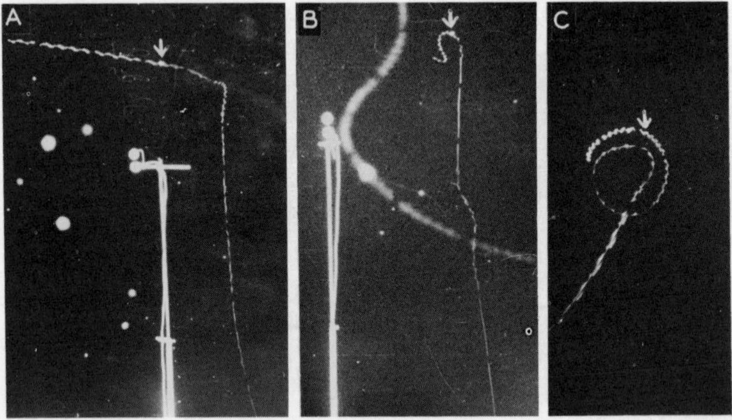

Fig. 17. Flight tracks made by unidentified moths in response to artificial ultrasonic pulse sequence (5-msec pulses of 75 kc/s recurring at the rate of 10 per second).[40] The multidirectional sound source is on a 14-foot mast. Gaps in tracks occur every 0.25 sec; arrows indicate a white dot marking the onset of the stimulus, which continues throughout each record; the end of each track is indicated where it passes out of the frame. The white dots and indistinct track in (*B*) were caused by other insects flying close to the camera. (*A*) Power dive; (*B*) passive drop broken by a short period of wing movement; (*C*) looping dive.

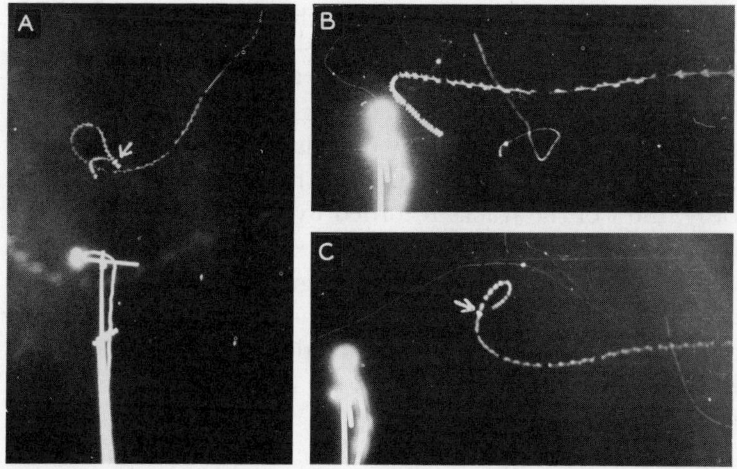

Fig. 18. Flight tracks made by unidentified moths showing directional responses to an ultrasonic pulse sequence.[40] Conditions similar to those of Fig. 17. The flare surrounding the sound source is due to overexposure, and does not indicate the presence of a light source. Other faint tracks and dots are caused by other smaller insects. (*A*) Oblique upward movement of a moth stimulated while over the sound source; (*B, C*) horizontal turns away from the sound source.

they could be captured, while others dived into the grass and vegetation and could not be found.

A few of the 1000 tracks recorded in this manner are shown in Fig. 17. Moths reacting within 10 feet or so of the transmitter show a bewildering variety of reactions usually ending in a dive, irrespective of whether the moth is above, below, or to one side of the transmitter at the time of the stimulus. The simplest reaction seems to be an abrupt dive with wings closed. This passive dive may be interrupted by short bursts of wing movement (*B*), or it may be a power dive (*A*). The dive may begin as an abrupt deviation from a horizontal flight path, or it may be preceded by a short climb, a loop, or a tight turn (*C*). Instead of a dive the moth may make a series of tight turns or loops that carry it more gradually to the ground. After landing, some individuals remain motionless for 30 seconds or longer, while others may resume level flight just before touching the grass.

These reactions lacked any obvious directional component, except that they always carried the moths downward. Naturally, most of the tracks recorded by the camera were made by moths flying close to the mast. At the same time, it was observed that some moths flying at greater distances from the transmitter would turn and continue in level flight directly away from the sound source. Reduction of the sound intensity brought these reactions closer to the mast, and it was possible to record a few tracks.

Some of these directional movements are shown in Fig. 18. They were sometimes preceded by one or more turns, and took the form of a straight or wavering path directly away from the transmitter. The direction taken appears to be influenced only by the relation of the moth to the transmitter, insects below taking a descending path, those at the same level a horizontal path, and in a few cases where moths were flying directly above the transmitter they turned and flew upward out of sight (*A*).

Thus, the prediction of the neurophysiological observations seems to be confirmed. High sound intensities produce nondirectional responses; low sound intensities produce directional responses. The great sensitivity of the tympanic organs must aid moths in moving out of the general area in which bats are feeding—the "early-warning" signal. As the number of impulses in the tympanic-nerve transmission increases to the saturation point this changes to the "take-cover" signal, and the moths dive for the ground.

Like most biological observations, this one raises a dozen questions for every one it answers. Most of the moths making these tracks certainly belonged to the families Noctuidae and Geometridae. How do the several families lacking tympanic organs, some containing common and successful species, survive without the ability to hear bat cries? Tympanic-nerve responses recorded from different noctuid species are generally consistent and similar, except perhaps for sensitivity. On the

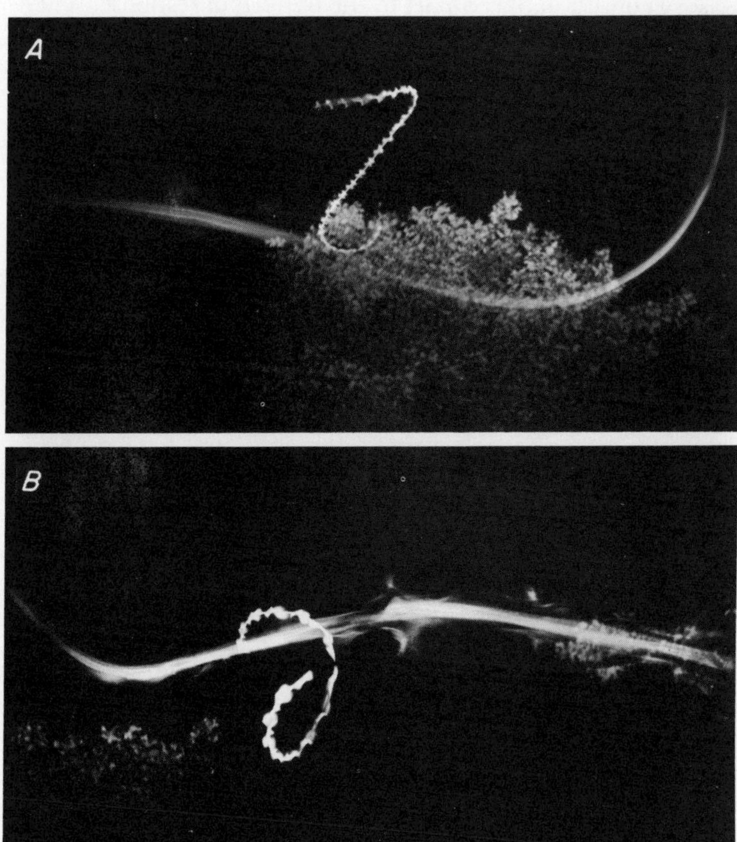

Fig. 19. Tracks of bats intersecting those made by moths tossed in the air. (*A*) The bat enters from the left; the moth dives and turns sharply, escaping capture. (*B*) The bat enters from the right; the moth track begins below that of the bat and terminates on it, indicating capture. (Photographed by Frederic A. Webster at Tyringham, Massachusetts.)

other hand, the variety of nondirectional maneuvers released by high-intensity ultrasonic stimulation defies any attempt at orderly description. Does each species have its characteristic pattern of response, or does it have a repertoire upon which it can draw in random order? Does sound intensity or some other sensory condition play a part in the pattern of response? There is some comfort in the thought that this unpredictability, however determined, is probably as confusing to the bats (Fig. 19) as it is to the experimenter, and therefore is of importance in the survival value of the behavior.

6. *Tactics for Two*

Before leaving bats and moths to their nocturnal game of hide-and-seek let us take a closer look at the problems of detection facing both contestants. The game takes place in darkness, and most of the moves—the ultrasonic signals and aerial maneuvers—require special instruments to reveal them to our senses. Therefore, any attempt to reconstruct the actions of the players is inevitably a string of speculations supported only at infrequent intervals by scraps of hard information.

In the preceding chapter a moth was regarded as an acoustic "point" having no dimensions. No attention was paid to the possible effects its body form might have on the signal echo received by the bat or on the moth's own capabilities of detecting its predator. A glance at a flying moth shows at once that its form is far from a simple geometrical figure such as a sphere. Furthermore, its shape changes from 10 to 40 or more times a second as it flaps its wings. Our visual image is certainly finer grained than the acoustic one reported by the bat's sonar, yet it seems likely that movement of the moth's wings has two effects on the interaction of moths and bats. First, the wing vibration must enrich the echo received by the bat. Second, it must also have profound effects on the acoustic sensitivity and directionality of the moth's ears, situated as they are just below the hind wings (Fig. 7).

Feeding bats seem to recognize the departure of their natural flying food from a passive spherical form. If one tries to fool wild bats as they swoop and turn in a natural feeding area by tossing moth-sized pebbles in the air, the hunters are seen to turn and follow only momentarily those objects as they sail past in a ballistic trajectory. Rarely if ever do bats actually attempt to intercept a pebble, even though it must present a simple problem in course prediction compared with that posed by an erratically flapping moth. One is forced to conclude that bats discriminate such objects from their natural food.

North American brown bats kept in captivity and fed mealworm larvae soon learn to catch such food if it is tossed to them while they are on the wing. Using high-speed photography and an ultrasonic sound track, Frederick Webster has revealed the details of the neat interception course plotted by bats with falling mealworms (Fig. 16). He has also been shown that bats become adept at discriminating falling mealworms from variously shaped but inedible objects such as spheres and spinning discs, apparently on the basis of the difference in their echoes.[15]

Echoes from Moths. Curiosity about the kind of acoustic profile a flying moth presents to a bat led me to construct a very simple working model of a bat's sonar system[41] (Fig. 20). A moth was fixed to a wire by means of a small drop of wax applied to its thorax. It was then provoked into stationary flight by lifting its feet out of contact with the ground, aided sometimes by the additional persuasion of a gentle head-on current of air. Directed at the flying moth from a distance of 50 cm were an ultrasonic loud-speaker and microphone placed below and above a prism. The loud-speaker emitted on demand single short ultrasonic pulses. The echoes of these pulses

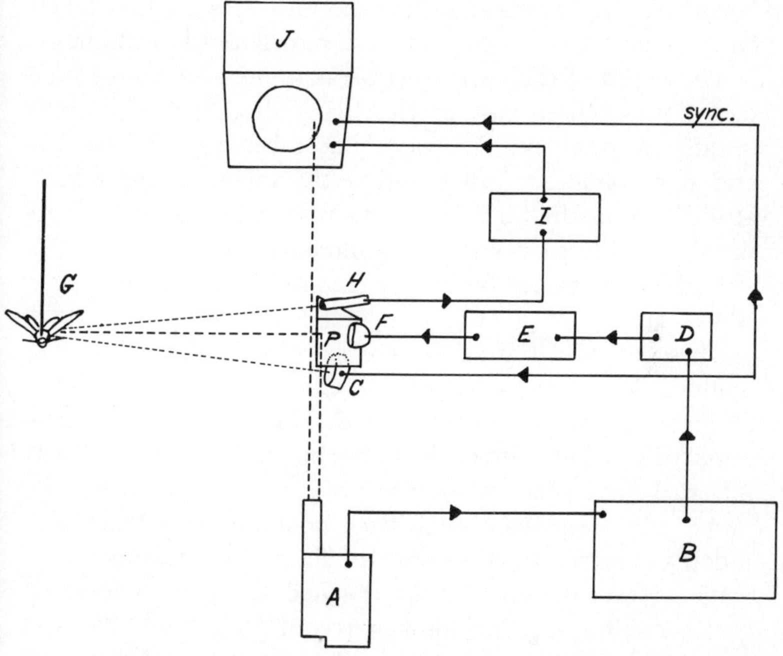

Fig. 20. Diagram of apparatus for comparing the ultrasonic echo from a flying moth with the instantaneous attitude of its wings. (*A*) Movie camera with built-in electrical contact that closes momentarily as shutter opens. (*B*) Ultrasonic pulse generator triggered by camera signal. (*C*) Ultrasonic loud-speaker. (*D*) Delay circuit. (*E*) Stroboscopic flash generator. (*F*) Strobe lamp. (*G*) Flying moth mounted on wire support. (*H*) Microphone. (*I*) Microphone amplifier. (*J*) Cathode-ray oscilloscope. (*P*) Prism. Pulse from the ultrasonic pulse generator passes to loud-speaker, to the delay and ultrasonic flash generator, and to the oscilloscope (*sync*) to initiate the horizontal transit of the beam. Delay is adjusted so that flash of light illuminates the moth at the instant the sound pulse strikes it. Prism serves to bring image of moth and oscilloscope screen onto each frame of film.[41]

from the moth's body and wings were picked up by the microphone and displayed as pulses on the oscilloscope screen. The horizontal sweep of the oscilloscope beam began on the departure of each outgoing pulse. Therefore, the horizontal displacement of the echo measures the time taken for the sound to reach the moth and return to the microphone.

The object of this experiment was to measure the relative size of the echoes cast as the flying moth held its wings in different positions. The prism placed between loud-speaker and microphone served to bring the image of the moth's attitude to one half of the camera field, while the other half simultaneously surveyed the oscilloscope screen. The moth's attitude was captured on film by the brief flash of an electronic strobe lamp.

These pieces of apparatus were coordinated in the following manner. A contact in the camera (either still or movie) closed momentarily as the shutter opened. This triggered the generation of a single ultrasonic pulse by the loud-speaker and initiated the horizontal sweep of the oscilloscope beam. The same signal triggered a single flash from the strobe lamp after a delay predetermined so that the flash of light would occur at the instant the sound pulse reached the moth. The sweep of the oscilloscope, including a record of the intensity and form of the echo, was recorded on the frame of film by the time the camera shutter closed. Thus, each frame included the optical and acoustic profiles of the moth in flight, both signals being registered for the same instant in time and from approximately the same angle.

From what has been said in the preceding chapter it may seem surprising that the moth continued to fly while being bombarded by high-intensity ultrasonic pulses. Indeed, many tested did not. Others could be persuaded to fly steadily only after repeated exposures, possibly due to adaptation of central nervous mechanisms concerned in the acoustic system. In some cases the problem was solved by destroying the tympanic membranes of the moth before the experiment.

It must be emphasized that this simple one-track sonar system is not meant to mimic the infinitely more complex one possessed by a living bat. Rather, it serves to transpose into

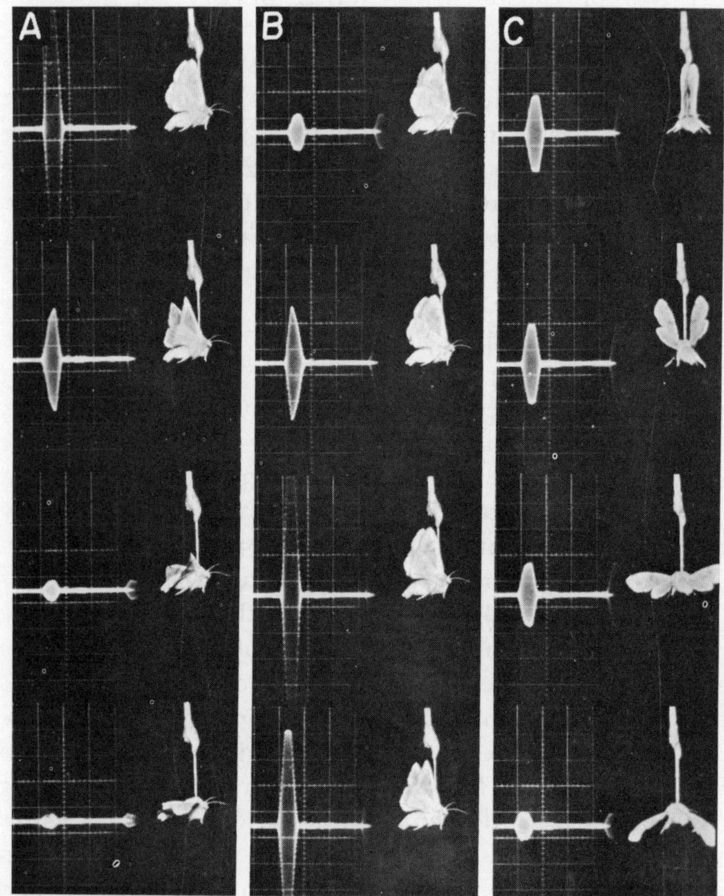

Fig. 21. Consecutive frames showing echo and attitude of a flying *Orthosia hibisci.* Camera speed (frames per second) was kept close to the wingbeat frequency of the moth so as to produce stroboscopic effect. (*A*) Lateral view during upper half of upstroke. (*B*) Several attitudes near top of stroke. (*C*) Rear view during upstroke.[41] Frame sequence reads from bottom to top.

terms appreciated by our senses the sort of signals reaching a bat's ear from a flying moth, and to compare these with the visible profile presented by the moth at the same instant.

Some of the pictures registered by this apparatus are shown in Fig. 21. In *A* are shown consecutive frames from movie film made as the moth, presented from one side, moved its wings from the middle to the top of their stroke. During this sequence the echo is seen to increase many times in size, although the visible profile of the moth changes relatively little. It is clear that the largest echo is cast back when the flat undersurface of the wings is momentarily at right angles to the sound path, and that the body of the moth is a relatively insignificant source of echoes. The exact angle of the wings appears to be very critical in producing a maximum echo. A stroboscopic trick was used—the frame speed of the movie camera being adjusted close to the wingbeat frequency of the moth (about 30 per second). By this means the flight movements were made to appear to stop at any position. In sequence *B* the camera shows almost identical wing positions, yet only in one are they in the correct position to throw back a maximum echo. When the moth was presented to the apparatus from other angles (sequence *C*) the echo also shows some fluctuation with wing position, but momentary presentation of the relatively large and flat undersurface of the wings occurs only when the moth's body axis is approximately at right-angles to the sound path. Of course, this statement is true only when the moth and sonar system are at the same altitude and the moth is flying "on an even keel." If the moth flies above or below the source of the beam of ultrasound directed at it, this flat presentation of the wing surface must occur at many other angles and relative flight paths.

The pictures suggest several possibilities. The acoustic representation of the moth lacks entirely the detail presented by the photograph, yet a comparison of several echoes indicates the wing position with great contrast. This means that to a bat the returning echoes must appear to have a pronounced "flicker" at the wingbeat frequency of its prey. Possibly this is how bats discriminate flying insects from tossed pebbles.

The maximum echo—perhaps 100-fold that coming from the body of the moth—occurs only when the wings are at right-angles to the sound path. If the contestants in this nocturnal game are flying at the same altitude with each on an even keel and straight flight path, this maximum echo will be detected by a bat at maximum range only at certain instants as the moth crosses its path at 90 degrees. But this is only one among many possible configurations. Field observation suggests that bats begin to chase moths only when they are 8 to 12 feet away. Perhaps this is the maximum distance at which they are able to receive a consistent echo rather than occasional "blips" from a moth-sized object that flaps its wings. On the other hand, a moth's ear begins to detect the cries made by a flying bat at a much greater distance (see below), when the bat's first indications must be quite intermittent or lacking altogether. At this range an echo can return to the bat's ears only when one of its sonar pulses happens to arrive at an instant when the moth's wings are in a favorable attitude. We have seen that one circumstance for this favorable attitude would be when the moth crosses the bat's path at right-angles. If the moth changes course on first detecting the bat and flies directly away from the sound source a favorable wing attitude will no longer be presented to the bat's sonar system. Such a change of course will cause the moth, from the bat's "viewpoint," to seem to disappear. This may be an additional justification for the turning-away behavior shown by moths exposed to faint ultrasonic pulses (Chapters 5 and 11).

On the other hand, a moth flying only a few meters from a bat makes nondirectional turns, loops, and dives (Fig. 17 and 19). At this range the bat probably receives an echo from its target at all wing positions. The moth cannot hide, acoustically speaking, by flying on any bearing relative to the bat. Nor is it likely to outdistance its pursuer by flying away from it. This seems to justify the change in the moth's behavior to nondirectional looping and diving when it encounters ultrasonic pulses at high intensity. At close quarters the moth's only defense is to make maneuvers that cannot be predicted or negotiated by the bat which, it must be remembered, does not get a continuous acoustic image of the moth but can only plot its interception course from a series of "dots" spaced by its own cries. During the final dogfight the bat decreases the spacing between these dots by increasing the repetition rate of its pulses and shortening their duration.

This attempt to reconstruct the maneuvers of bat and moth is grossly oversimplified and no doubt very incomplete. First, it assumes that both contestants are flying at the same altitude and on smooth courses. Actually, a moth's progress is marked by yawing and pitching (Fig. 22) as well as by major swoops and turns. Second, the artificial sonar system employed a pulse of constant pitch, whereas the cries made by North American brown bats contain frequency modulation. This must complicate the echo pattern returning from a flying moth. Third, moths flap their wings at a fairly constant frequency, though it varies with species from 10 to over 40 per second. A cruising bat emits its cries about 10 times a second, increasing to over 100 per second when a source of echoes is detected nearby. It is difficult to say how these two frequencies —the bat's pulses and the moth's wingbeat—interact in determining the echo pattern received by the bat's ears. It is clear that there is much to be learned before we can appreciate the finer points of the game.

The Range of a Moth's Bat Detector. At this point it is necessary to answer a question that has so far been sidestepped. How sensitive is a moth's ear? Like most simple questions this one is not easy to answer in a meaningful way. First, one must choose some criterion of response by the more sensitive *A* cell that might have significance in terms of a moth's behavioral reaction. A spike pattern intermediate between those shown in the 5-decibel and 10-decibel frames of Fig. 10 was chosen for reasons that will be given in Chapter 11.

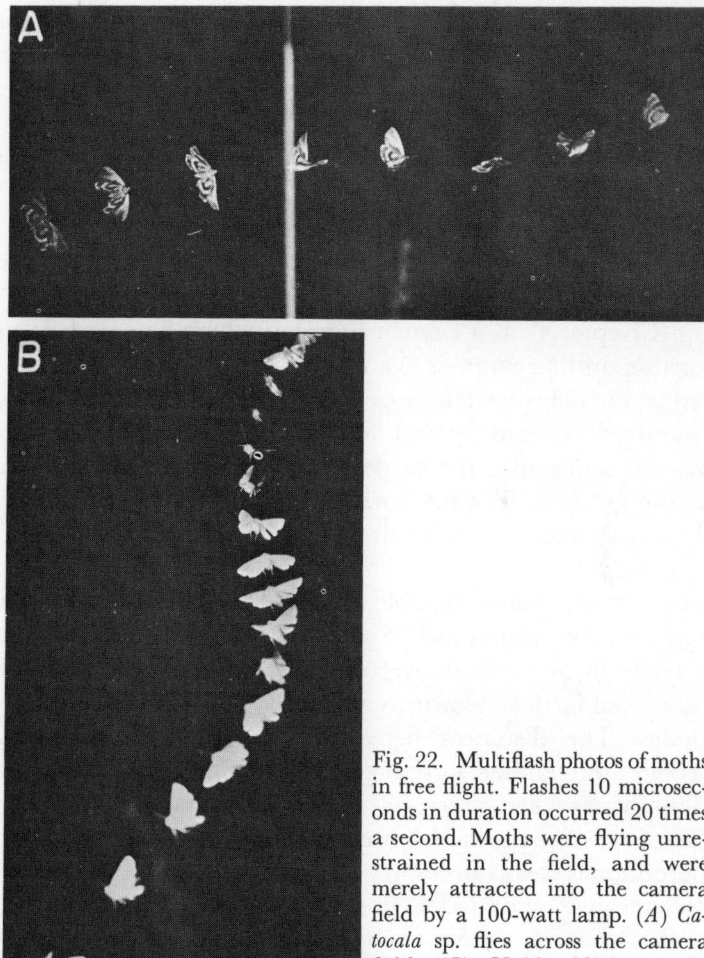

Fig. 22. Multiflash photos of moths in free flight. Flashes 10 microseconds in duration occurred 20 times a second. Moths were flying unrestrained in the field, and were merely attracted into the camera field by a 100-watt lamp. (*A*) *Catocala* sp. flies across the camera field. (*B*) Unidentified noctuid flies down and towards the camera.

Second, one must measure the sound pressure necessary to give this response and express it in physical language—in dynes per square centimeter. Measurement in these terms is essential if the biophysical properties of moth ears are to be compared with the properties of other animals' ears, but it can be argued with equal force that moths don't know anything about dynes, and that in this chapter we are concerned with the tactics of moths and bats, not with biophysics. Therefore, the question of sensitivity must also be asked in another way. What is the maximum distance or range at which the cries of a cruising bat produce this criterion *A* response in a moth's ear?

An approximate answer has been found to both questions.[43] That to the first is 0.01 to 0.03 dyne/cm^2. The answer to the second was obtained by setting up a tympanic nerve preparation (Chapter 4) in a Concord garden where bats were wont to cruise on fine summer evenings. The situation is diagrammed in Fig. 23. Bats customarily left their roost in a red barn a hundred yards away and flew singly or in pairs in a fairly constant course over the garden on their way to their evening feeding grounds. The territory was open and the bats generally took a straight path, all the while emitting their sonar chirps.

Two microphones capable of detecting these chirps were placed on the ground in-line and "upstream" on the bats' expected path, and a moth prepared for recording the *A* response was placed farther "downstream" and in line with the microphones. The distances between these three bat-detecting stations were chosen so that the bat was in range of the two microphones when its cries were just detected by the moth ear. Therefore, since the moth ear is more sensitive than man-made microphones it was necessary to place it a great deal farther down the bats' expected flight path. Ground distances between the three stations were measured, and the signals coming from them were registered on three channels of a high-speed tape recorder.

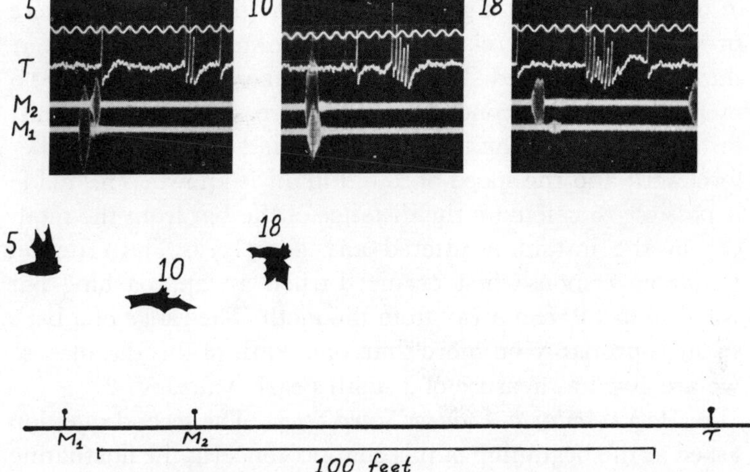

Fig. 23. Responses of the tympanic organ of *Catocala* sp. to the cries of a bat approaching in the field. The upper frames show the tympanic responses (T) to the 5th, 10th, and 18th pulses made by the bat during this approach. The same pulses are also registered on each frame by the microphones M_1 and M_2. Time scale at the top of each frame is 100 c/s. Scale diagram below shows the three detectors arranged in line and roughly along the bat's approach path. Distance scale is shown below the baseline. Silhouettes (not drawn to scale) show the distance of the bat when each cry was emitted.[43]

A great many passes of bats were recorded as they flew over the array. Most of the records had to be discarded because the bats failed to fly a straight course or swerved at a critical moment. If more than two bats were in range at the same time the record became too complex to analyze, and the number of usable records was further reduced by failure of the tympanic organ, turbulence in the air, or a sudden rain shower. Photographic recordings made from tapes of the remainder showed signals registered at the three stations by each chirp as the bat passed down the line (Fig. 23).

Sound travels in air at about 1160 feet per second. Therefore, the first detector to register a given chirp was that closest to the bat as it passed overhead, and the relative differences in time of arrival of a given chirp at the three detectors were proportional to the relative distances from them of the bat at the instant it chirped. These relative times of arrival could be measured in milliseconds from the record; the linear (ground) distances between the three detectors had been measured beforehand; and the speed of sound in air is known. This made it possible to calculate the distance of the bat from the moth ear at the instant it uttered any given cry. The criterion tympanic response first occurred when an approaching bat was 100 to 140 feet away from the moth. The range of a bat's sonar is probably no more than one tenth of this distance, so we are given a measure of a moth's early warning.

A Moth's Ability to Locate a Sound Source. The second question asked at the beginning of this chapter concerns the fluctuating shape of a moth as it flaps its wings. Do the wings, hinged as they are just above the opening of the ear, act as rapidly moving acoustic baffles? This would cause fluctuations in the directional sensitivity of its ears during flight.

An answer to this question requires measurement of the relative sensitivity of the moth ear to sounds coming from different directions and when the wings are in different positions. Roger Payne, of Tufts University, assisted by Joshua Wallman, then an undergraduate at Harvard College, undertook to make these measurements.[29] Electrodes were to be placed under the tympanic nerve of a moth and connected to electronic equipment for amplifying and displaying the spike potentials transmitted from the acoustic receptors (Chapter 4). A loud-speaker generating artificial ultrasonic pulses at a fixed distance from the ear was to be moved to various angles relative to the body axis of the moth, and the intensity of ultrasound needed to produce an arbitrarily chosen but fixed tympanic response was to be measured at each position. The reciprocal of the ultrasonic pulse intensity needed at each point would give a relative measure of the sensitivity of the ear to sound presented to the moth from that angle.

Stated thus in outline, the experiment appears to be quite straightforward. One is tempted to connect up the necessary apparatus and gather the data before morning. Before examining just how Payne and Wallman obtained the measurements, let us take a more thoughtful look at the theoretical and practical problems to be faced in planning an experiment of this kind.

Ideally, we would like to have this information from a moth freely flapping its erratic course through the night. In other words, we would like to be unobtrusive observers in the moth's world of darkness, sound, and movement. But we are brought somewhat rudely back to our own world by the hard fact that the only way to measure the tympanic response is to attach to the tympanic nerve two hundred or more pounds of electronic gear, for no telemetering amplifier has yet been developed that is quite small enough to strap on a moth's back. A second requirement of the plan also forces us to use a moth fixed in space. Measurements must be made one at a time with sound transmitter and moth receiver in fixed positions, and they must be compared with one another. Therefore, the response must be at a measured and fixed distance from the sound source.

Even though these requirements dictate that the moth must be stationary, it is still possible to mount it thus and allow it to flap its wings in some approximation of their natural rhythm (Fig. 27). But here we are forced to make a second compromise with the natural situation. We wish to know the relative acoustic sensitivity of one ear to sound coming from all points in an imaginary sphere having at its center the moth with its wings in a certain attitude of flight. A moth normally flaps its wings many times in a second. Not only would it be difficult to collect all these measurements during that small fraction of a second when the wings are in a certain phase of their beat, but the surgery needed to expose and record from the tympanic nerve requires destruction of much of the flight muscle. Even if an operation could be devised that permitted the moth to flap its wings during the experiment, the violent vibrations accompanying this activity would probably obscure and distort the relatively small electrical signals coming from the tympanic nerve. Therefore, the moth must be stationary and unable to flap its wings, and we must fix them in different postures, preferably like those assumed in normal flight, as we make our measurements.

There are still other problems to be faced in planning the experiment. We wish to make a long series of measurements for comparison. But in exposing the tympanic nerve we have destroyed much of the moth's tissue and interfered with its respiratory and circulatory systems. This inevitably places the subject on the downgrade predicted by "times arrow" (Chapter 1). What if the tympanic sensitivity should change or the tympanic organ cease to function before a series of measurements had been completed? The first of these possibilities was allowed for by making the system of data collection as rapid and automatic as possible, and by making frequent replicate measurements to check for changes in tympanic sensitivity. The second caused many uncompleted runs to be discarded.

These methodological defects in the acoustic localization experiment have been enlarged upon, not because it is a bad experiment, but because they are examples of the sort of conditions and limitations imposed in most biological experiments by the experimenter and his artifacts on the natural circumstances of his experimental animal. Much is made of objectivity in science, and some researchers are so misguided as to think they can achieve it. Experimental biologists are generally modest in this respect, perhaps because of a feeling of some commonality with their experimental animals. The best a biologist can hope for is awareness of his own subjectivity, that is, the degree to which he commits himself to become part of the experiment when he imposes his conditions on the animal and causes it to depart from its natural behavior. Subjectivity intrudes itself into biological research at many other points. It is impossible to begin an experiment without a hypothesis, yet this already commits the researcher to a degree of "tunnel vision" or partiality. We have seen how subjectivity enters into the planning of the experiment, that is, in deciding how to become involved with the animal. And, when the experiment is completed, the experimenter is obligated to use some measure of imagination in his attempts to extrapolate from the mutilated specimen in his apparatus to other members of its species free and intact in their own world, and to visualize their existence when they are endowed with senses or powers first revealed by his experiment. Perhaps this last step is the ultimate in subjectivity, yet without it the experiment has no meaning.

But let us return to the preparations in the laboratory. Payne and Wallman first built a chamber out of material that would minimize echoes (Fig. 24). In its floor was a revolving platform surmounted by a tower of slender wires. The apex of the tower was in the center of the chamber, and consisted of a trident of fine needles on which the body of the moth was firmly impaled. The needles served as an indifferent electrode. The active electrode on which the tympanic nerve was looped was a fine platinum hook insulated except at its tip and connected to a preamplifier below. The platform and tower with the moth mounted at its peak could be rotated smoothly through 360° while tympanic nerve impulses were being recorded without interruption.

An ultrasonic loud-speaker, 90 cm from the moth, was mounted on a vertically hinged arm so that it could be moved to all positions from directly above the moth to a point almost beneath it, remaining throughout at a constant distance and directed at the preparation. The loud-speaker

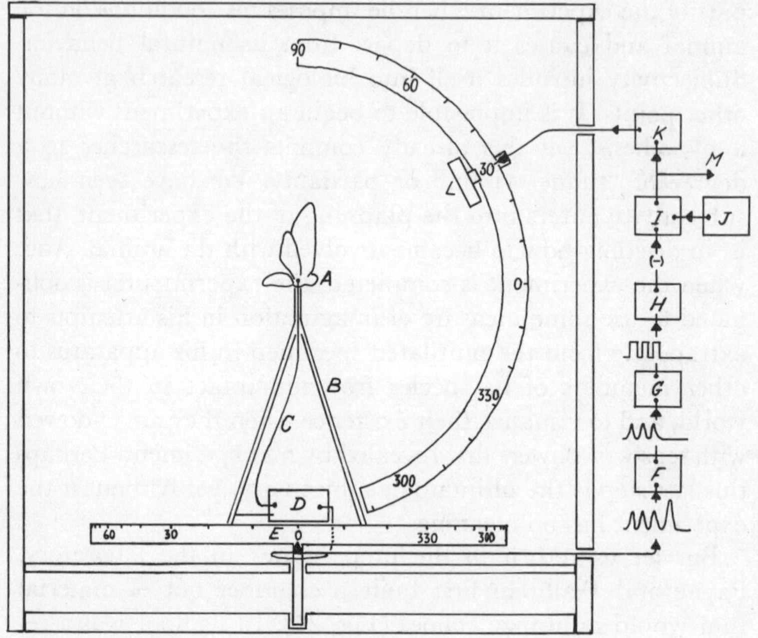

Fig. 24. Schematic diagram of apparatus for measuring the directionality of a moth's ear. Moth (*A*) with wings fixed in position is mounted on tower (*B*) which serves as indifferent electrode. Active electrode on one tympanic nerve is connected (*C*) to preamplifier (*D*). Tower and preamplifier stand on graduated turntable (*E*). Cable from preamplifier carries amplified tympanic nerve spikes outside anechoic chamber to pulse-height window (*F*) which separates *A1* spikes from other impulses picked up from nerve. *A1* spikes are converted to uniform pulses by (*G*), and in turn to a proportional voltage by (*H*). This serves as the negative feedback signal acting through the feedback amplifier (*I*) to regulate the output of the ultrasonic pulse generator (*J*). The train of regulated ultrasonic pulses is amplified by driver (*K*) and returned to the anechoic chamber to be emitted as sound by loud-speaker (*L*) on graduated track. The amplitude of regulated ultrasonic signal is automatically recorded (*M*) as turntable is rotated. The position shown for the moth and loud-speaker (indicated by black arrows) corresponds to 30° vertical and 0° horizontal on the Mercator projections.[29]

arm was provided with a scale marked in degrees of elevation, and the platform had a scale marking the angle between the body axis of the moth and the loud-speaker arm.

The plan was to measure tympanic sensitivity as the moth was rotated through 360° with the speaker at a certain angle of altitude. The loud-speaker was then moved to another altitude and the process repeated until all altitudes (in 10° steps) had been covered above and below the horizontal. Measurements could not be made directly below the moth because at this point the loud-speaker arm collided with the tower. This meant that the ear was eventually presented with sound pulses coming from most points on an imaginary sphere surrounding the moth.

One run, the collection of data for one sphere, meant making about 500 measurements. Since it was intended to collect several spheres of data from the same preparation, each with the wings of the moth in a different position, a formidable number of readings had to be made. Although some tympanic nerve preparations lasted for 8 hours, one even for 14 hours, hand collection of the readings on sensitivity would have been much too time-consuming. So an automatic method was devised for measuring the sensitivity of the tympanic organ.

Pulses of 60 kc/s sound and 6 msec duration were emitted 20 times a second by the loud-speaker. For reasons that will be apparent later it was desired to measure the directionality of the ear to faint sounds—only a few decibels above its threshold. Accordingly, a response of between 2 and 3 spikes per sound pulse in the most sensitive *A* receptor was chosen as criterion. This response level is close to that shown in the 5 decibel frame of Fig. 10. Each group of *A* spikes was amplified, electronically separated from random *B* spikes, and converted to uniform pulses of constant height but similar in number and spacing to the original *A* impulses. This signal was then converted into a voltage whose height was determined by the number of pulses—and allowed to operate a feedback amplifier that regulated the intensity of the signal emitted by the loud-speaker. This amplifier operated on a negative control principle similar to that regulating constant length in the muscles of vertebrates (Chapter 8) or steering the strike of the praying mantis (Chapter 12). An increase in the number of *A* spikes caused a larger signal to enter the feedback amplifier which then decreased the intensity of the ultrasonic pulses emitted by the loud-speaker; only two *A* spikes caused a smaller feedback signal and less attenuation of the sound. Thus, a constant *A* response was maintained, regardless of direction of the stimulus, because the nerve response continually adjusted the stimulus intensity.

This made the process of data collection automatic and continuous. All one had to do was to record on a strip chart the intensity (voltage) of the signal going to the loud-speaker as the platform bearing the moth was steadily rotated. This voltage was inversely proportional to the sensitivity of the moth's ear. It was plotted on the vertical axis of the chart, while the moth's position relative to the loud-speaker was entered on the horizontal axis.

In spite of this time-saving device there was much to be done. Thirty or more of these charts were needed, each at a different elevation of the loud-speaker, in order to construct a sphere of relative sensitivity readings for one moth with its wings in one position. Sometimes only one or two spheres could be obtained before the tympanic organ died. However, a number of times 4 or 5 spheres were obtained, and in one case 11, from individual moths.

The problem of deciding how to analyze the mass of data collected in these experiments was almost as perplexing as that encountered in designing the method of collection. We wished to visualize the acoustic realm presented to a moth's ears, that is, to know how sounds register on one ear in terms of their relative intensity when they come from many different directions. In order to understand the method chosen for analysis it is helpful to picture a moth at the center of an imaginary globe (Fig. 25). It is headed directly for the 0°

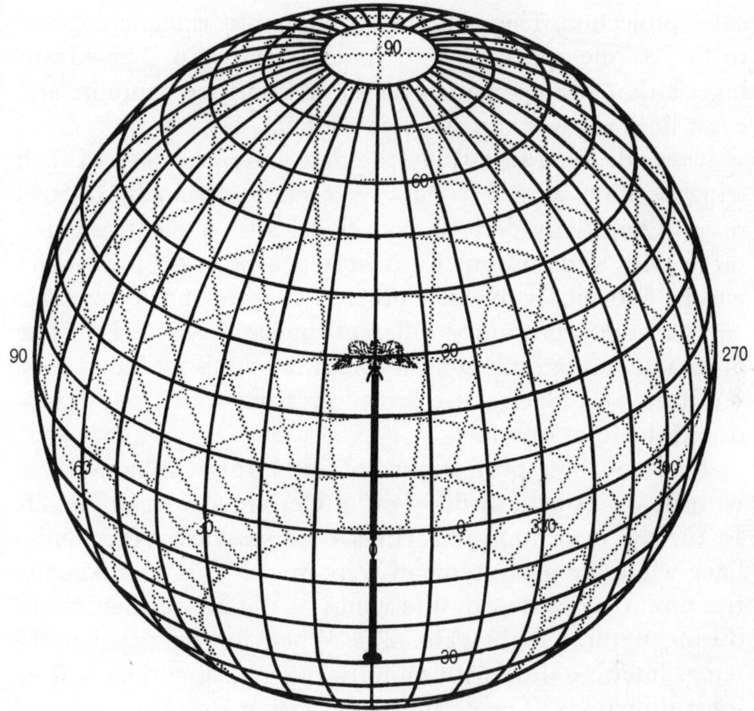

Fig. 25. Moth pictured at the center of an imaginary globe. Measurements of the relative acoustic sensitivity of one ear were made from the directions indicated by the intersection points of the latitude and longitude lines, the whole set was repeated for each wing position. These measurements formed the basis for making the Mercator projections (Fig. 26). (Courtesy *Scientific American.*)

longitude line, and the horizontal plane of its body coincides with the equator. This fixes the moth's axes in a set of three-dimensional polar coordinates.

The problem of presenting features on the surface of a sphere has already been faced by the global mapmaker when he is allowed only the two dimensions of a sheet of paper. He compromises with three-dimensional reality by using a Mercator projection. The cost of this compromise is increasing distortion as one approaches the poles of the map. The advantage is that the eye can survey and compare contours and coast lines all over the globe at a glance.

It was decided to display the data for each moth at each wing position as a Mercator projection. Decibel values of relative sensitivity derived from the charts were entered on a large grid, and contour lines were drawn to enclose values within 5 decibels of each other. Areas bounded by these contour lines were tinted different shades of gray, with white areas indicating directions of sound to which the ear is most sensitive, and black those directions from which sound produced the least response.

Three of these projections obtained from the same moth with its wings held at different angles are shown in Fig. 26. In the top projection the wings were fixed over the moth's back while the measurements were made. This approximates the position reached by the wings at the top of their stroke during natural flight (Fig. 27). When in this position the wings interfere little with sound waves reaching the ear from most directions. The contours show that there is a marked right–left asymmetry in acoustic sensitivity, the right ear detecting sounds most effectively when they originate in the horizontal plane on the right side and roughly at right-angles to the body axis. The region for minimal hearing for the right ear is the corresponding zone on the left side of the moth.

When the wings are moved to positions approximating those assumed near the middle and bottom of the wing stroke (middle and lower projections) there is a change in this acoustic asymmetry. For each ear the zone of greatest hearing deficit moves from the opposite side to an area directly above the body of the moth, and there is a slight improvement in sensitivity to sounds coming from directly below. Thus, the wings appear to act as moving acoustic baffles, replacing

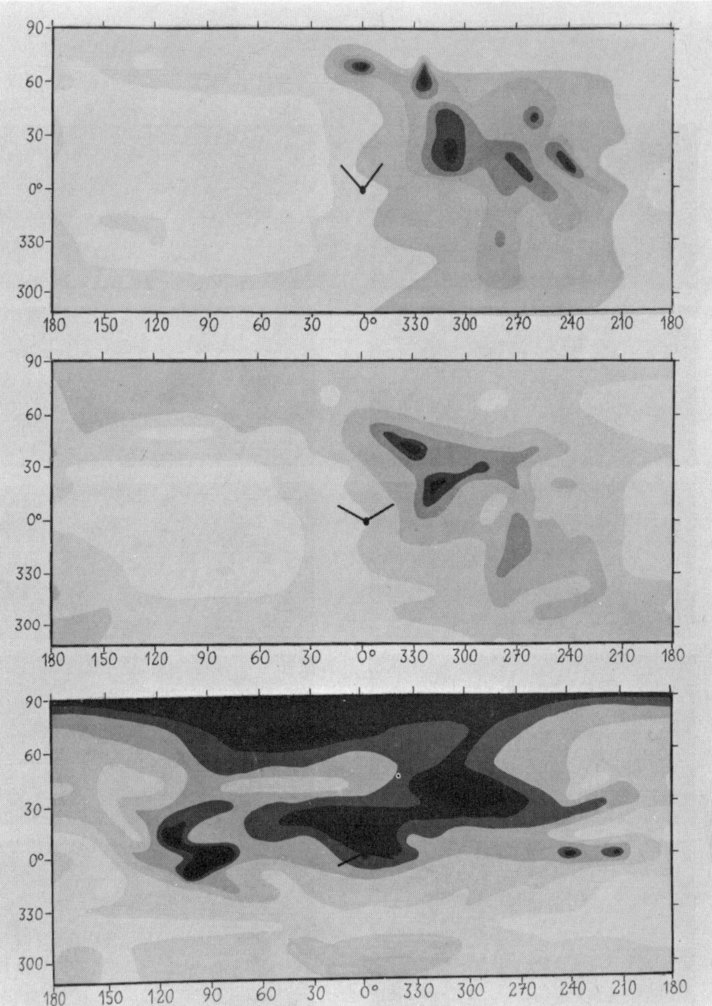

Fig. 26. Mercator projections showing directional sensitivity of the right tympanic organ of *Catocala unijuga* when the wings are held at three different angles. Profile of the moth giving wing position is shown headed toward the viewer and located at 0° on the vertical and horizontal scales. Test pulses were 60 kcps. Shades of the contours are in 5 decibel steps, light tinted areas indicating directions to which the right tympanic organ is most sensitive, and black those directions to which the ear is least sensitive.[29]

Fig. 27. Strobe-flash pictures of the wing positions assumed by *Caenurgina erechtea* while in stationary flight. (*A*) Anterior view. One complete wing cycle is shown beginning with the top of the downstroke. (*B*) The same viewed from the left side. From this angle the opening of the tympanic organ is clearly visible in the top and bottom frames. Compare with Figure 7.

the right–left acoustic asymmetry existing when they are above the back with a dorsal–ventral acoustic asymmetry as they sweep down and around the body.

This is the main change imposed by the moving wings on the sound field of the ear. If one disregards the fine detail shown by the projections and allows for the distortion at their poles the only other noteworthy point is that directly ahead of and just below the moth's body axis there is a narrow zone that is but little affected by wing position. Directly behind the moth there is a similar but much broader zone of shallow acoustic contours that is likewise relatively unchanging as the wings move.

This brings us once more to that point in experimental biology where we must attempt to withdraw ourselves and our machinery from the life of a living creature and try to visualize it going about its natural business. Our objective has been to bring to light a new facet of its existence, but to achieve this we have had to remove the subject far from its natural context and to expose it to many constraints and irreversible insults. Starting from this point, theoretical reconstruction of the *status quo* is necessarily a highly subjective undertaking on our part.

We have revealed that right–left acoustic asymmetry alternates with dorsal–ventral asymmetry as the wings of a moth are moved from the top to the bottom of their stroke. What can this mean in the life of a moth? If we fail to make an attempt, however tenuous, to answer this question the time of the investigator and the lives of many subjects have been wasted.

Commonly, moths flap their wings from 10 to 40 times a second, and with a sculling motion that is much more complex than the attitudes in which they were fixed for the experiments on directionality. Actual attitudes assumed by the wings in stationary flight are shown in Fig. 27. There is little information about these attitudes in free flight (Fig. 22), so we must assume that they are not very different. As the wings begin their downward sweep (Fig. 27) the camera has a clear lateral view (*B*) of the tympanic opening below the hindwing. The camera view is approximately from the region of greatest tympanic sensitivity of the near-side ear. In later stages of the downstroke the tympanic opening is obscured. The pictures show that the natural movement is much more complex than the simple flapping we tried to mimic in the directional experiments. Complex changes in the angle of attack of the wings at the bottom of the downstroke and the beginning of the upstroke suggest changes in the directional sensitivity of the ears at which we can only guess.

A second assumption we are forced to make is that a moth travels in a straight line. A few moments observation shows that this is far from the case. In addition to major turns and loops, there is minor stalling, yawing, and pitching in its progress. Also, the movement of the wings must cause the body of the moth to oscillate rapidly up and down. The long axis of the body may approach the line of flight when a moth is flying rapidly (Fig. 22), but it tilts up when the moth is flying slowly or hovering. All of these variables must be considered in trying to answer the question raised above.

During natural flight right–left acoustic asymmetry must alternate with dorsal–ventral asymmetry 30 or 40 times a second. This fact may help explain how a moth is able to plot a course directly away from a source of faint ultrasonic pulses, even though their intensity is sufficient to stimulate only the most sensitive *A* receptor in each ear. Evidence was presented in Chapter 5 that the differential in the response of right and left *A* receptors contains sufficient impulse-coded information to enable the moth to steer a left or right turn away from a distant source of pulses. But this alone does not tell us how a moth could discriminate a sound coming from above from one coming from below, and thereby be able to select a flight path downwards or upwards and away from the source. Yet, free-flying moths are observed to do this with some accuracy.

A moth needs another piece of information beyond that supplied by the right and left tympanic organs in order to steer away from a sound source first detected at any angle to its body axis. This could be the instantaneous position of its own wings. The Mercator projections suggest that the moth could be informed that the sound source was above it if the signal transmitted to its central nervous system from the *A* receptors "flickered" from "strong" to "weak" as the wings passed through their downstroke, and then reversed this order during the upstroke. A reversed relation between wing phase and strong–weak fluctuations in the tympanic nerve signals might indicate a source below the moth. As suggested in Chapter 5, difference between right and left *A* receptor signals, measured by the central nervous system while the wings are elevated, would locate the source to the right or left. Information on wing position could, of course, come from the motor neurons supplying the flight muscles or from receptors detecting movement of the wings or thorax. Some evidence of this mechanism is given in Chapter 11.

In considering this interpretation it is hard to avoid the assumption that the moth sails along on a straight course like an airplane. Yet, as has been mentioned, this is generally not the case. Various perturbations in its progress must appear to cause the distant sound source to swing through many bearings relative to its body axis during each second of flight. During the first few moments of response to a distant source of pulses there is erratic "hunting" behavior (Chapter 5), although the course finally assumed when flying away from the source may be relatively straight. This suggests a slightly different view of the steering mechanism, although it is based on the same two sets of signals—those from the more sensitive

A receptors in the two ears, and from some indicators of its own wing position.

It has already been pointed out that there are two zones relative to the moth's body axis from which a sound impinging on the tympanic organs will not "flicker" with wing angle and will produce little or no right–left asymmetry. These are a small zone directly ahead of the moth and a large zone directly behind. The area ahead subtends a very small solid angle, and it seems likely that the erratic path of a moth first encountering sound coming from this direction would soon bring the bearing of the source into one of the lateral zones of maximum "flicker" and differential between the right and left tympanic nerve messages. This might initiate random "hunting" movements—turns, climbs, and dives—rather than specific reactions to commands from the ears and sensors of wing position. If these continued until the moth received signals that showed neither right–left asymmetry nor modulation by its own wingbeat, in spite of the normal perturbations associated with its own progress, the sound source would be directly behind it. The moth would then be on course directly away from the source.

The experiments described in this chapter suggest that the form of the moth and the movement of its wings may add various dimensions to the contest between bat and moth. The fluctuating echoes thrown back by the moth's wing movement may aid the bat in recognizing its prey. At the same time they may enable a distant moth to disappear, acoustically speaking, by turning away so as not to present the flat surface of the wings to the bat's sonar beam. If the moving wings cast echoes they can also act as acoustic baffles. This quality may provide the moth with another sensory dimension, which, when coupled to the information coming from the right and left *A* receptors, could enable it to plot a course away from a distant bat approaching from any angle.

Each of the contestants in this nocturnal war game has certain tactical advantages. The bat has speed and power; the moth, with its smaller bulk, has greater maneuverability. The bat probably does not get a steady signal from its prey until within 8 to 12 feet; the moth appears to have the capability of detecting its predator at ten times this distance. In the light of this the two forms of behavior shown by the moth—nondirectional dives, loops, and turns on receiving intense ultrasonic pulses, and turning and flying away from a source of faint signals—seem to have survival value.

A Moth Answers Back. Other gambits in this nocturnal game for survival await detailed study. Three British biologists—A. D. Blest, T. S. Collett, and J. D. Pye—found that certain arctiid moths generate ultrasonic clicks when they are handled.[3] Their noise-maker consists of a series of minute corrugations or microtymbals in the cuticle forming the basal segment of the third pair of legs (Fig. 28*A*). When the muscles below are contracted and relaxed the cuticle is bent and

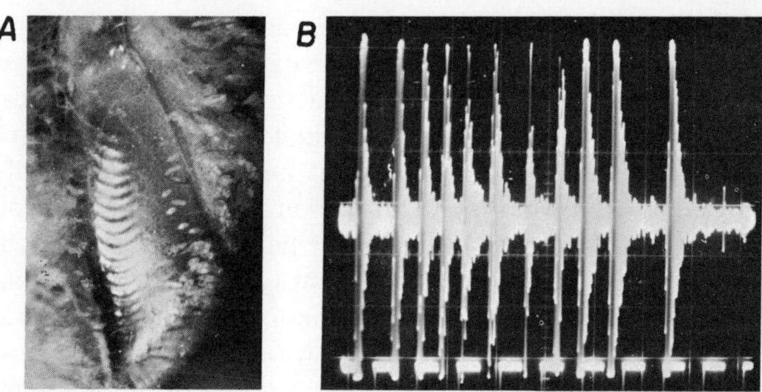

Fig. 28. The click-producing organ on the metathoracic episternite of *Halicidota tessellaris*. The scales have been brushed away and the microtymbals show up as highlights on the smooth cuticle. (*B*) A series of clicks produced by the microtymbals of this species in response to simulated bat cries. The broken line marks millisecond intervals.[9]

unbent and these corrugations "click" in sequence, each after the manner of a party clicker. Thus, the moth is able to produce several thousand brief pulses of ultrasonic noise a second (Fig. 28*B*).

Arctiid moths are closely related to noctuids and have similar ears. Dr. Dorothy Dunning and I found that certain New England arctiid species could be induced to turn on their clickers if they were exposed to batlike sounds while suspended and in stationary flight.[9] Since bats hunt their prey on the basis of acoustic signals this behavior looks as if it should provide extra guidance for the predator. Such altruism is not to be expected in nature, and Dr. Dunning searched for another explanation of the arctiids' behavior. She found that New England bats refuse to eat certain arctiid species, whether they are presented alive or dead. It appears that the moths' noise-making when a bat approaches advertizes their distastefulness and thus aids in their survival. From a bat's viewpoint a prospective meal that answers back is likely to taste bad.

This is as far as it is possible to go at present in examining the signals, moves, and tactics of the game for survival played by moths and bats. The nerve signal generated by the acoustic receptors has been repeatedly used as an index of the information available to a moth, but we have been primarily concerned with what the moth does or is capable of doing on receipt of this information. At this point it is reasonable to ask about the neural machinery that transforms the *A* signal into avoidance behavior. What goes on in a moth's central nervous system when its ears are stimulated by a bat's cry? The following chapters explore very different methods and concepts in order to form a basis for understanding the behavior and interactions of neurons within the insect central nervous system. We shall return to the ganglia of a moth in Chapter 11.

7. *Evasive Behavior in the Cockroach*

Many tricks are used by animals in evading their predators. If action occurs it is usually fast, for a few milliseconds may distinguish the quick from the dead. This promises simplicity in the neural machinery of evasive behavior. The present chapter is concerned with some general aspects of evasive behavior, and with a closer look at some of the neural mechanisms in another example.

Evasive behavior differs from the behavior patterns with which ethologists are generally concerned because the whole action is directed toward breaking off the interaction between stimulus source and subject. In its extreme form—the startle response—there is little or no feedback even from the beginning of the action. The ducking, freezing, or wild flight caused in many animals by a single loud sound or a sudden movement continues long after the termination of the stimulus, and its direction, if any, is influenced little, if at all, by the characteristics of the stimulus. The relation between stimulus and response can be said to be "open," or lacking in feedback (Fig. 29*A*). Action is initiated by the stimulus, but its direction and duration are determined by the subject.

Startle responses of this sort grade into steered avoidance behavior such as that described in the previous chapter. Here

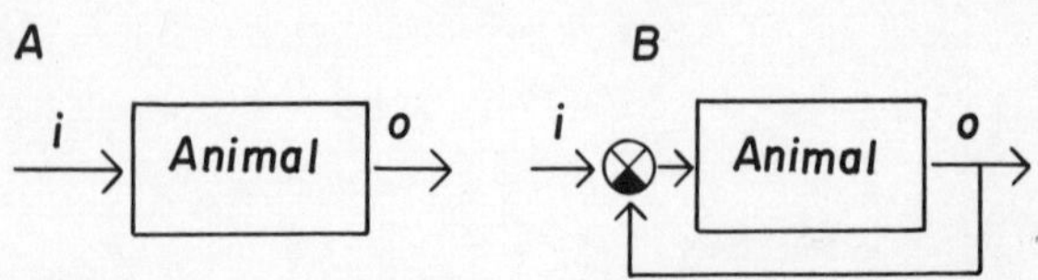

Fig. 29. "Open" and "closed" interaction of an animal with the environment. (*A*) The stimulus *i* causes a response *o* that is not steered through feedback. This open interaction is characteristic of many types of evasive or startle response. (*B*) The response continuously influences the animal's relation to the stimulus through negative feedback (black segment in circle). This interaction is characteristic of most types of continuous orientation.

the coupling between the stimulus source and the subject remains "closed" while the latter moves away, but the nature of the subject's action is such as to open it once more. An obvious and important aspect of evasive behavior is that the action invariably overshoots or continues for some time after contact with the stimulus source has ceased. It could be said that the optimum stimulus strength for the subject is zero.

The majority of ethologically interesting behavioral situations are "closed." The stimulus contains elements causing both approach and withdrawal of the subject, the optimum stimulus situation having some fixed or fluctuating value other than zero. This is true of orientation to certain nonliving aspects of the environment, such as steering by physical or chemical gradients, as well as of conspecific interactions such as social relations, courtship, and care of the young. Stimulus source and subject remain coupled by a closed feedback loop (Fig. 29*B*)—in conspecific situations each individual is at the same time stimulus and subject—and the behavior continues until it leads to a consummatory action or is broken off for some other reason.

Another factor that might be expected to make it relatively easier to work out the neural mechanisms of evasive behavior and its complement, predatory attack, is the degree to which the success of both parties depends upon a short response time. This must have been a potent factor in the evolution of attack and escape mechanisms. If it is assumed that the populations of prey and predator have remained roughly balanced over a number of generations, then each genetic modification making possible a reduction in the response time of the predator must have spread rapidly through the predator population, and greatly increased the selection pressure acting on the prey population for a decrease in the response time for escape, maneuver, or concealment. Any randomly occurring genetic change favoring this decrease might be expected to spread through the prey population. This reciprocal selection pressure to shorten the attack and evasion responses must have pushed the evolution of the relevant neural mechanisms to a level of efficiency and simplicity limited only by other biological needs and by the basic limitations of neural mechanisms. Even though in a behavioral sense the action tends to uncouple the predator and prey as rapidly as possible, in the evolutionary sense the contestants are irrevocably and tightly coupled. Indeed, this interaction probably goes much further than a single prey-predator pair. In the long and complex food chains found in nature the predator of one species is usually the prey of another, so that the modifications evolved in connection with one role may influence the efficiency of the other.

It is worth while to pause and consider the advisability of selecting a single factor out of the network of influences regulating an animal's life before using the factor of response time as a lever in trying to pry out some more information about the neural mechanisms of behavior. This emphasis is liable to the same sources of error as those surrounding the measurement of the selective advantage of evasive behavior in moths (page 63). In considering one instant and one experience in an animal's life there is no way of weighing its value against other experiences and needs, both individual and racial, such as, for example, the fact that the males of many spiders and insects approach the female during mating, only to be killed and eaten during the process (Chapter 10). The *reductio ad absurdum* of this emphasis would be the evolution of an animal having organs for predator evasion to the exclusion of reproductive organs.

Nevertheless, like increasing size, increasing speed of movement and reaction show up as evolutionary trends in most animal phyla. Unlike size, speed has unfortunately left no direct geological traces, but among living animals those species of a group thought to be closer to the ancestral stem are frequently both smaller and slower in their actions.

Giant Fiber Systems. More direct evidence that speed in predator evasion has influenced the evolution of the nervous system is to be found in the giant fiber systems of worms, squid, crustacea, and insects. These fibers are axons of relatively enormous diameter occurring both within the central nervous system and as motor axons supplying certain muscles. They are not equally developed in all species of these groups, but they occur sufficiently often, and always in connection with mechanisms of withdrawal or escape, to emphasize the value of the velocity of nerve impulse conduction in surviving attack by a predator.

It is interesting to attempt an estimate of the cost and of the contribution of increased axon diameter in this context. A cross-sectional view of the giant axons in one half of the ventral nerve cord of the cockroach is seen in Fig. 2. The ventral nerve cord is the main trunk of fibers connecting the paired ganglia of each body segment, and thus constitutes the insect's central nervous system through which must run all nerve fibers connecting anterior and posterior regions of the body. At the level shown in the section it can be seen that most of the axons are less than 5 microns in diameter, although 6 percent of the area is occupied by three axons 25 to 30 microns in diameter. Each of these giant fibers has an area roughly equivalent to that occupied by 100 fibers each 3 microns in diameter. The giants conduct impulses at a velocity of 6 to 7 meters per second, or in about 2.8 milliseconds over their full length. Conduction velocity in 3-micron fibers of the cockroach nerve cord has not been measured, but other studies indicate that it is probably somewhat less than one-tenth of that of the giants.

Disregarding the time element, a given sequence of nerve impulses in a large axon transmits the same amount of information as a similar sequence in a small axon. However, the information-handling capacity of 100 small axons operating in various numbers and combinations is astronomically greater than that of a single large axon, even if it takes ten times as long for this information to begin arriving at the other end of the fibers. In the evolution of the cockroach's nervous system this additional capacity for handling information must have had less survival value than the few milliseconds saved in transmission time by the presence of the giants. The same principle is evident in the warning systems developed by man. It is much more important that a danger signal should consist of rapidly transmitted information such as "Fire!" or "Take cover!" than that it should be delayed in order to transmit all the details of the threat.

A Startled Cockroach. The giant fibers shown in Fig. 2 are internuncial, that is, they lie entirely within the central nervous system, and serve to connect sense cells on the cerci (Fig. 31) to motor centers that innervate the leg muscles. The behavior with which the giant fibers are concerned is readily observed by anyone rejoicing in daily contact with cockroaches, particularly the large and active American cockroach. At night, when cockroaches are most active, the observer should slowly approach a single insect standing motionless near the center of an unobstructed area such as a wall or floor. A short puff of air directed at the cerci, the small antennalike structures on the tip of its abdomen, will send the roach scurrying off and probably out of sight.

The response time of this startle reaction was measured in the following way (Fig. 30).[38] A cockroach was attached to a thin balsa-wood support by means of a small amount of hot wax applied to its pronotum (the broad dorsal shield just behind the head). The other end of the support was placed in the stylus holder of an ordinary phonograph pickup. The feet of the cockroach were allowed to grasp an unattached piece of cork or balsa that rolled round and round as the insect made walking movements. A second phonograph pickup bearing a small paper flag was mounted near, but not touching, one of the roach's cerci, and a small tube was adjusted so that

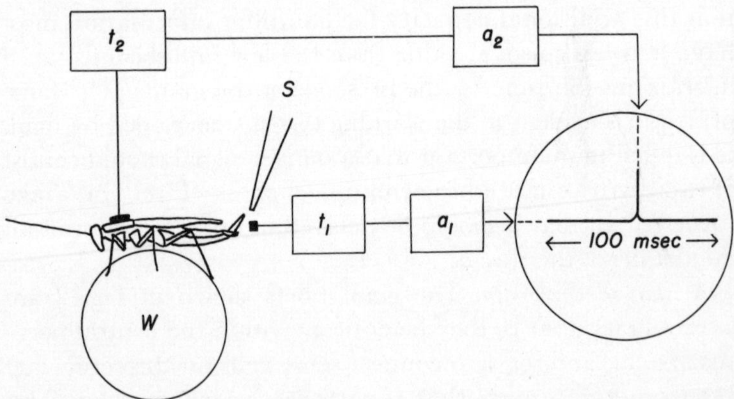

Fig. 30. Diagram of the method for measuring the startle-response time of a cockroach. Transducer t_1 detects air movement from a jet *s* near the cercus of the cockroach, and triggers the horizontal sweep of the oscilloscope through amplifier a_1. Transducer t_2 detects the ballistic thrust of the cockroach as it begins to run on the ball, *W,* and indicates through amplifier a_2 the initial movement as a pulse on the oscilloscope.

a sudden jet of air could be applied to the cercus and the paper flag at the same instant.

The outputs of both pickups were amplified and connected to a cathode-ray oscilloscope so that a pulse from the "puff detector" triggered the horizontal sweep deflection of the beam, while a pulse from the "movement detector" caused a vertical deflection on the ensuing horizontal trace. The velocity of the horizontal movement of the beam was preadjusted and measured, and a camera was arranged so as to record the signal.

Patience was needed with the subjects, since many continued to walk, clean themselves, or make other movements that caused vibrations in the pickup. Others repeatedly kicked away their "walking platforms." When a cockroach had been motionless for some time, the camera shutter was opened and a brief jet of air was squirted on the cercus. This bent the cercal hairs and simultaneously caused a pulse in the nearby pickup that started the oscilloscope sweep. The first movement of the startled roach was a vigorous forward leap that kicked the platform out behind. The corresponding forward-directed recoil of the insect's body deflected the support pickup and registered as a pulse on the calibrated trace.

Twenty-three measurements gave response times for the startle reaction ranging from 28 to 90 milliseconds, with an average value of 54 milliseconds. The method could have been responsible for little variation in these times, so the range must be due to variation in events occurring within the cockroach. A constant but small fraction of this response time is the 2.8 milliseconds needed for transmission along the giant fibers. Before attempting to make a temporal balance sheet for the startle response by adding up the durations of the various neural events intervening between stimulus and response, it will be necessary to glance at the structures concerned.

Nerve Pathways in the Startle Response. The central nervous system of a cockroach is diagramed in Fig. 31. Each ganglion is connected by nerves with the sense organs and muscles of its body segment, its size corresponding to the local number and complexity of these organs. The ganglia belonging to some segments have become fused together, particularly those at either end of the body, and all are joined by paired connectives into a nerve cord, most of which lies close to the ventral side of the body. The brain is dorsal to the digestive tract, and joins the ventral nerve cord through connectives passing on either side of the esophagus. It consists of three ganglia more or less fused into a single mass, and is primarily concerned with antennal and visual senses.

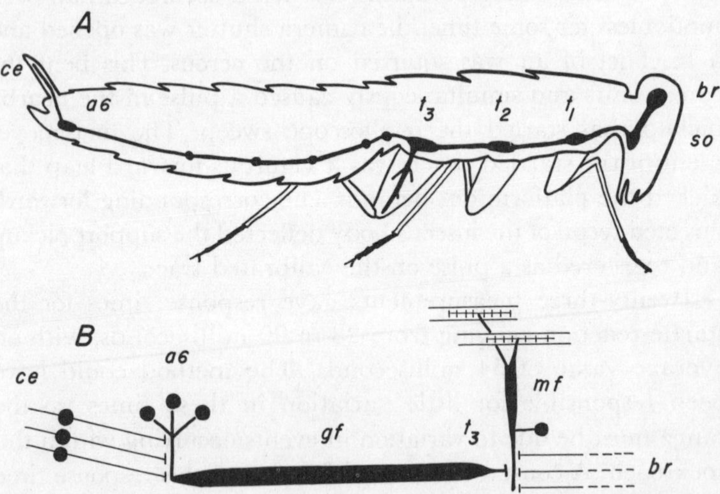

Fig. 31. (*A*) Diagram of the central nervous system of the cockroach, showing the nerve supply to the cercus and the metathoracic leg. (*B*) Diagram of the main nerve elements concerned in the startle response. *a6,* last abdominal ganglion containing synapses between afferent fibers from cercus and giant fibers; *br,* brain; *ce,* cercus; *gf,* giant fibers in abdominal nerve cord; *mf,* motor neuron supplying leg muscles; *so,* subesophageal ganglion; *t1, t2, t3,* thoracic ganglia.

The section of the nerve cord directly concerned with the startle response is shown in Fig. 31*B*. The cercus is covered with fine hairlike sensilla, each delicately poised in a small socket in the cuticle. The sensilla are deflected in their sockets by gentle air currents and even by low-pitched sounds. This displacement excites sense cells lying just below the cuticle. The sense-cell axons, about 150 in number, form the cercal nerve that enters the last abdominal ganglion of the nerve cord. It will be seen that this sensory arrangement is much more complex than the two-fiber system of the moth's ear, so that the pattern of afferent nerve impulses cannot be decoded with the same ease. However, the elongated arrangement and greater size of the cockroach's nervous system compared with the small compact pterothoracic ganglion of the moth make it much easier to experiment with the central connections made by the sensory fibers.

Having entered the last abdominal ganglion the 150 cercal nerve fibers make contact or synapse with about four of the ascending giant fibers and an unknown number of smaller axons. It requires about 2.0 milliseconds after bending of the hairs for cercal-nerve impulses to reach this point. Little is known of the anatomy of these synapses except that, although the cercal axons converge closely enough on the giants to affect them, there is no anatomical or functional continuity between the two sets of fibers. Synaptic discontinuities of this sort are thought to make up the primary discriminating mechanisms of the central nervous system. The general importance and principles of synaptic transmission will be considered in the following chapter, and for the moment we shall be concerned merely with its influence on the startle response.

The nature of this influence becomes clearer if the cercal nerves are stimulated with electric shocks regulated so as to control both the number of cercal fibers active and the frequency of cercal impulses arriving at the synaptic contacts with the giant fibers. The response of the giants to this cercal bombardment is detected by recording their spike potentials from electrodes placed further up the nerve cord.[34,36]

The manner in which impulses in the cercal fibers generate impulses in the giants at these synapses is distinct from the process of self-propagation of impulses in axons (Chapter 3) in the following ways: (1) the cercal or giant axons can transmit 200 to 300 impulses per second for extended periods without fatigue, but the synaptic process fails and transmission is blocked after a few seconds of bombardment by presynaptic impulses at this frequency; (2) various anesthetics and drugs as well as other chemicals affect synaptic transmission long before they affect the transmission of impulses in axons; (3) synaptic transmission is one-way, taking place only from cercal nerves to giants, although both types of axon can transmit impulses equally well in either direction; (4) there is a hiatus or delay between the arrival of cercal-nerve impulses at the synaptic region and the departure of giant-fiber impulses (Fig. 32). The synaptic delay is 1.1 to 1.5 milliseconds, being longest when only a few cercal fibers are stimulated; (5) the giant fibers cannot be made to discharge unless impulses in a certain *number* of cercal nerve fibers arrive more or less simultaneously at the synaptic regions (Fig. 32). Impulses in one or a few cercal fibers produce no detectable effect on the giants, but it follows that they must have had some local effect within the ganglion because if a few more impulses in other fibers overlap their arrival a full-blown giant spike is generated. Since the cercal fibers are assumed each to have a separate point of contact as they converge upon a giant, this overlapping or additive effect of spatially separated local actions is called spatial summation.

In passing it is worth noting that the synaptic properties listed above all have a negative or subtractive effect upon the spread of impulses through the nervous system. Each imposes

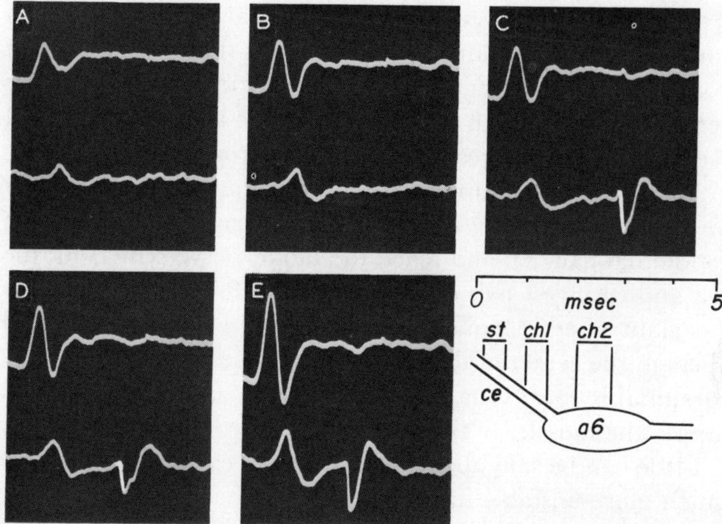

Fig. 32. Responses recorded from the surface of the cercal nerve (*upper traces, ch1*) and the terminal abdominal ganglion (*lower traces, ch2*) of the cockroach following electric shocks of increasing strength (*A* through *E*) applied to the cercal nerve (*st*).[36] The diphasic compound action potential in the cercal nerve (*upper traces*) becomes progressively larger as the increasing stimulus excites a greater number of cercal-nerve fibers. The cercal-nerve potential is also evident as the first upward deflection in the lower traces. Spatial summation at the giant-fiber synapses is sufficient to excite one giant fiber in (*C*) (downward deflection), and two giants in (*D*) and (*E*). Decreasing synaptic delay with increasing afferent volley is evident in (*C*), (*D*), and (*E*). Last abdominal ganglion, *a6;* Cercal nerve, *ce.*

some condition or limitation of the probability of an impulse being generated in a giant fiber. They add up to a discriminatory process analogous to that occurring in a sense cell, which is excited only by certain modes of external change. Likewise, the postsynaptic response is generated by incident presynaptic impulses arriving only under certain conditions, and the generator process is irreversible and marked by a finite delay. Behavior seems to be the outcome of a string of discriminatory events, and is itself a discrimination—a noncontraction of many specific muscles that distinguishes it from a convulsion. Further speculation of this sort must be deferred to Chapter 8 while we continue to trace the neural events of the startle response of the cockroach.

The giant-fiber impulses sweep up the abdominal nerve cord, possibly slowing momentarily as they pass through the abdominal ganglia, and reach the motor nerve centers for the last and strongest pair of legs in about 2.8 milliseconds. Here the giant fibers narrow and disappear in a tangle of small fibers in the center of the methathoracic ganglion, where they presumably form synapses with motor neurons whose axons supply the muscles of the leg.

Little can be said about these synapses except that they are much more unstable in their operation than those formed by the cercal-nerve fibers and giant neurons.[34,36] Frequently the incident impulses produced by electrical stimulation of the giants failed to generate impulses in the motor neurons after one or two trials, and the whole system often appeared to block merely because of the dissection needed to expose the nerves and ganglion, or because of the restraint needed to hold the insect motionless during the experiment.

After a number of frustrating attempts enough was learned about these interactions between giant fibers and motor neurons to add two more synaptic properties to the list compiled for the cercal-nerve–giant-fiber synapses. First, postsynaptic impulses were propagated only after two or more successive volleys of presynaptic impulses had arrived. Often, the response did not occur until after the third or fourth presynaptic volley. This can be explained only by assuming that the early "ineffective" volleys did indeed leave some trace that faded with time. This trace played a part in the generation of a postsynaptic response only when it was added to similar traces left by later volleys traveling the same path. This addition of local actions in time at synapses is known as temporal summation. Second, once a sufficient number of volleys had arrived to generate postsynaptic impulses in the motor fibers, the latter continued to repeat for several seconds after the incoming volleys had ceased. This afterdischarge is analogous to that occurring in the moth's *A* fibers when the ear had been stimulated by an intense but brief ultrasonic pulse.

The instability of this synapse makes it the weak link, and hence a determining factor in the startle response. In view of this importance it is tantalizing that the same properties make experimentation so difficult and uncertain.[32] It is not even possible to be sure that only one synapse is involved, because the temporal summation makes it difficult to measure the synaptic delay. Assuming a synaptic delay of 2 milliseconds, and allowing for the fact that a minimum of two giant fiber volleys separated by a minimum interval of 2 or 3 milliseconds must arrive at the synaptic region before the first motor impulse is able to depart, a minimum of 4 to 5 milliseconds, and possibly twice as much time, must be allowed for synaptic events at the metathoracic ganglion.

The final events of the startle response take place outside the nerve cord, and are measured with greater ease. An electric shock is given to a motor nerve as close as possible to the metathoracic ganglion. An electrode inserted into one of the main extensor muscles of the leg detects the arrival of the motor spike potential, and the development of excitation in the muscle fibers is shown by the slower and large muscle potential. Finally, the first movement of the leg is registered by a phonograph pickup or other electromechanical transducer.

It is interesting to compare the sum of the durations of the separate neural events as determined by these physiological measurements with the startle-response times measured earlier in intact cockroaches. But it must be remembered that for the physiological measurements only the shortest possible durations have been given. Thus, both of the synaptic delays are often longer, the first by about 0.5 millisecond, and the second possibly by several milliseconds. Also, it is difficult to estimate how long it takes the most time-consuming event—development of tension in the muscles—to set the cockroach in motion. Allowing for this the comparison is reasonably close (times in milliseconds): (*a*) Duration of neural events in the startle response: response time of cercal sense cells, about 0.5; conduction time in cercal-nerve fibers, 1.5; synaptic delay in last abdominal ganglion, 1.1; conduction time in giant fibers, 2.8; synaptic delay in metathoracic ganglion, 4.0; conduction time in motor fibers, 1.5; neuromuscular delay and muscle potential, 4.0; development of contraction, 4.0; total, 19.4; (*b*) Behaviorally measured response time, 28 to 90. (All measurements were made at temperatures of 22 to 25°C.)

It is tempting to play further with these figures, but hardly warranted because of the uncertainties mentioned above. With regard to the totals, it is comforting to find that the minimal physiological measurements, totaling about 20 milliseconds, are smaller than the shortest behaviorally determined startle-response time. The reverse would have been embarrassing!

The figures likely to show the least variation and error in measurement are those for conduction times in axons. If these are added separately they total 5.8 milliseconds. Thus about 10 percent of the average response time of 54 milliseconds, or 20 percent of the shortest, is taken up by the transmission of impulses along axons. This relatively large percentage adds weight to the anatomical evidence, mentioned earlier in this chapter, that there has been selection pressure toward the evolution of large axons in evasive mechanisms. An increase of 10 percent in conduction velocity attained by an increase in axon diameter or by other means would shorten by 1 to 2 percent the startle-response time of the cockroach, giving a selective advantage of significant magnitude in species evolution.

Startle Times in Other Animals. It is worth while comparing these response times with a few others before turning aside from the topic of speed in predator evasion. The natural balance assumed to exist between the population density of a predator and that of its prey indicates that the contest is not one-sided, that is, that it has the characteristics of a game. It would be interesting to have an analysis of the rules and odds for both sides in a specific predator-prey contest, but as far as I know this has never been made. Even figures for the strike times and startle times of unrelated predators and prey are very scattered. The elegant studies made by Mittelstaedt[27] of the factors steering the strike of the praying mantis should be followed up by a study of the evasive behavior and the startle times of its natural prey made so far as possible under field conditions. The mantis strikes from ambush, that is, it remains motionless and presumably undetected until the attack is begun, so that its first movement is the signal for the prey to take evasive action (p. 212). The strike takes 50 to 70 milliseconds to complete (Fig. 33). Laboratory-raised mantids fed on blowflies miss their prey in about 10 to 15 percent of their strikes, but it is not clear whether this is due to chance or to evasive action by the prey. Anyone who has tried to capture flies by hand knows that their startle-response time taxes the human strike time. A few measurements of the time required by flies to commence flying after their feet have lost contact with the ground gave values of 45 to 65 milliseconds, similar to the startle-response times of cockroaches. If either insect were struck at by a mantis, and was alerted by the first movement of the predator, it might just be beginning to move when hit. The response times of moths to ultrasonic stimuli are somewhat longer—75 to 252 milliseconds with a mean of

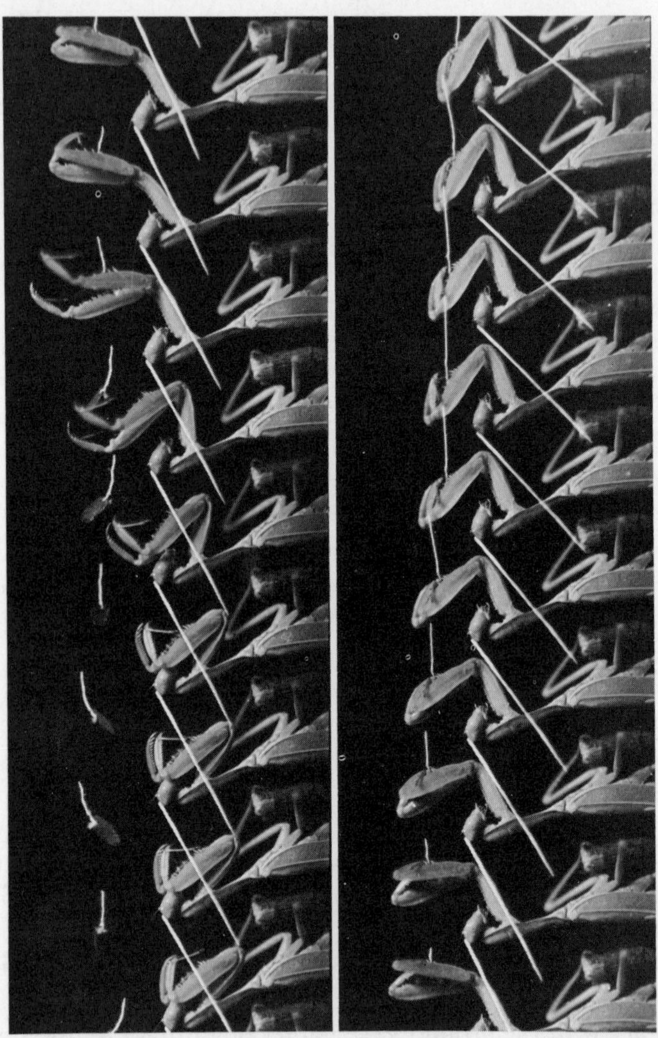

Fig. 33. An adult female mantis (*Heirodula* sp.) strikes at and catches a twirling paper lure.[31] Photographed on continuously moving film by electronic flash at 10-msec intervals. The sequence begins at bottom left. The mantis views the black lure against a white background. The camera's view shows the white string bearing the black lure against the black background required by the photographic method. The white line is a balsa-wood marker attached to the insect's head for another experiment.

139 and 143 milliseconds for two species in tethered flight.[60] However, more precise laboratory measurements of the turning-away tendency of flying moths when exposed to ultrasonic pulses coming from one side gave response times of about 45 milliseconds (Chapter 11).

Before leaving the startle response of the cockroach it is worth while to extract two more pieces of information out of the comparison of neural and behavioral events. About one-quarter of the time occupied by the neural events is taken up by the two synaptic delays. The methods used in these experiments were unable to detect any on-going activity during these synaptic intervals, yet it is apparent that both synaptic processes, particularly the second, are of the greatest importance in determining the nature of the evasive behavior. This becomes apparent if two more aspects of the evasive behavior are noted.

If we return to the initial experiment of puffing air onto a cockroach as it rests undisturbed in the center of a large open space, it will be seen that it responds less energetically to a second puff, and if this is continued several times in succession the stimuli ultimately elicit only short jumps or possibly no reaction at all. Other kinds of stimulation show that muscular fatigue is not the primary cause of this failure to respond, and one can only conclude that the cockroach has adapted or become used to the stimulus. If it is left undisturbed for some time, the response returns as strongly as ever.

A search for the physiological site of this adaptation can be made by exposing each of the neural components of the startle mechanism to repeated stimulation. This shows that the sensory mechanism of the cercal sensilla and the cercal, giant, and motor axons continue to operate throughout prolonged stimulation, the synapses between the cercal fibers and the giants are less stable, and the synapses between the giants and the motor neurons fail after only a few repeated volleys.

This marks the last as the controlling factor in this aspect of the behavior.

Another obvious and important characteristic of this and other types of startle behavior is that the response continues for some time after the stimulus has ceased to act. The cockroach runs for several feet upon receiving a single puff of air, and usually disappears into a crevice unless this is prevented. Avoidance responses would have little protective value if they ceased as soon as they had carried the animal out of range of the stimulus. In the cockroach this behavioral overshoot seems to be connected with the afterdischarge of motor impulses following stimulation of the giant fibers. Thus, the giant-fiber–motor-neuron synapses are a determining factor also in this aspect of the startle response.

The causes of this afterdischarge are obscure, and probably several in number. It may be due in part to an intrinsic instability in the mechanism of impulse generation, as in the *A* cells of the moth. Impulses from other parts of the nervous system undoubtedly play a part. During movement there must be continuous feedback from mechanoreceptors on the legs. In the cockroach the arrival of impulses from the brain seem to play an important part. A decapitated cockroach lives for several days, but it can be made to respond with difficulty, and then only with short jumps, to puffs of air on the cerci. In a nerve preparation the afterdischarge in the motor neurons following giant-fiber stimulation is reduced or absent after the head has been removed.

These casual observations make it obvious that the discriminatory processes occurring at points of synaptic contact between neurons are of supreme importance in the neural mechanisms of behavior. The following chapter will outline what has been discovered about the intimate details of the synaptic process through inserting microelectrodes into individual neurons in the mammalian spinal cord.

8. *Discrimination*

This chapter heading was chosen because it has wider implications than terms such as sensory transduction and synaptic transmission. A lock discriminates between different keys, an automatic telephone exchange discriminates between different number combinations, a sense cell discriminates between different stimulus modes, and a neuron in the central nervous system discriminates between different combinations in time and space of incident presynaptic impulses. In each case the outcome of the discrimination is quite simple: the presence or absence of movement of the lock, of the ringing of a certain telephone bell, and of a propagated impulse in the axon of the sense cell or of the central neuron. The input mechanism presents the problems.

An axon can communicate, but it is a poor discriminator, for it can be made to discharge identical impulses by a variety of different stimuli (Chapter 3). Discrimination depends upon negative properties, that is, upon the ability not to respond to many kinds of stimuli. Professor Norbert Wiener has pointed out that the value of a telephone exchange lies not in that it connects you with the correct number, but rather in that it does not connect you with wrong numbers. It would be much simpler to eliminate the telephone exchange and connect all the subscribers in a common net, but the results would be chaotic.

A specific behavior pattern in an animal could be described with equal accuracy in terms of what the animal does not do. Most insects and vertebrates have between 200 and 300 separate muscles. A behaviorally significant movement, that is, an action that promotes the survival of the individual or the species, is brought about by the contraction of a select fraction of these muscles in precise relation to one another in time and space. An equally active and significant part of the behavior pattern is the noncontraction or relaxation of the rest. This nonresponsive or inhibitory component of behavior is not necessarily determined by anatomical barriers or by the lack of certain connections between receptors and effectors, for under some circumstances the same stimulus may evoke conflicting behavior patterns. In the courtship of many animals sign stimuli produced by one sex may have an ambivalent effect on the opposite sex, releasing behavior that contains components of both approach and retreat. However, these opposite actions can only alternate, since if they were to occur simultaneously the result would be a convulsion.

The importance of inhibition in behavior is revealed by a well-known experiment with the convulsant drug strychnine. The physiological action of strychnine is to block certain inhibitory mechanisms in the spinal cord of vertebrates. In a strychninized frog stimuli that normally release adaptive behavior, such as hopping, cleaning, or withdrawal of the leg from a noxious stimulus, now cause simultaneous contraction of every muscle in the body. Stimuli through the medium of any of the sense organs has the same effect. All the subscribers are now connected to a common net so that a call put in at any point rings all.

Discrimination obviously begins with the sense organs. As far as we know at present, evasive diving in moths is released by ultrasonic, but not by visual or olfactory, stimuli. Other animals may show evasive behavior to visual or olfactory stimuli but not to ultrasound. Little is known about the mechanisms that endow each group of sense cells with a fixed capacity to discriminate one type of external change from all others. It is achieved partly by modifications that shield the receptive surface from all except one mode of stimulation. Some of these are obvious; for instance, the location of the acoustic cells does not favor stimulation by light, nor are the retinal cells, lying deep in the fluids of the eye, easily reached by air-borne vibrations. Many are much less obvious, such as the mechanisms distinguishing olfactory from taste receptors, both of which respond to chemicals. In addition to shielding, many sense cells contain mechanisms that enormously enhance their special sensitivity compared with that of undifferentiated nerve cells. An example is the retinal pigments in the visual sense cells of vertebrates and other animals. How these amplifying mechanisms serve to generate nerve impulses is still a mystery.

The relatively fixed discriminations provided by the sense organs are of less interest to us here than the more labile discriminations made within the central nervous system. The response of the *A* cells to ultrasound is relatively stable and reproducible in the same moth, and similar in different individuals and species of noctuid moth, yet the behavior engendered by the arrival of *A* impulses in the central nervous system is varied and unpredictable. The startle response in the cockroach wanes readily on repeated stimulation and varies in different individuals, yet responses in the cercal-nerve fibers and the giant fibers remain stable over long periods of repeated stimulation.

The relative simplicity in insects of these and other input mechanisms is an open invitation to investigate the subsequent discriminatory events within the insect central nervous system. Some of the difficulties in doing this are discussed in the following chapter, but the main reason that there has not been more progress is that few attempts have been made with sufficient precision and intensity. Neurophysiological tools are available, but they have not been put to use. The following pages provide an example of what has been learned from a much more complex system.

Postural Mechanisms in Mammals. Our present knowledge of discriminatory mechanisms in the central nervous system is based mainly upon a long series of experiments on the neural basis of postural reflexes in mammals. The impressive roster of contributors ranges from Sir Charles Sherrington to Sir John Eccles, whose technique of inserting glass microelectrodes into individual neurons lying deep within the spinal cord is providing most of our current information.[10,52]

The experiments of Eccles and his associates have been focused primarily on some simple but important components of the mechanism that regulates extension and flexion of the hind legs in mammals. They were carried out mainly on cats, but the mechanisms are undoubtedly very similar in man.

Steady standing requires that the leg be kept extended, and that the hip, knee, and ankle joints be stiffened. This is brought about by contraction of the extensor muscle groups at each of these joints. Most of the experiments have been concerned with the muscles of the knee joint (Fig. 34). The increase in tension in these muscles can be felt by placing one's hand on the front part of the thigh while rising from a sitting to a standing position. The tension increases and then remains constant as one continues to stand.

Tension in each extensor muscle is maintained by a continuous barrage of motor-nerve impulses originating in a "pool" of hundreds of motor neurons in the spinal cord. An axon from each motor neuron in the pool goes to a specific group of 10 to 100 muscle fibers among the many thousands

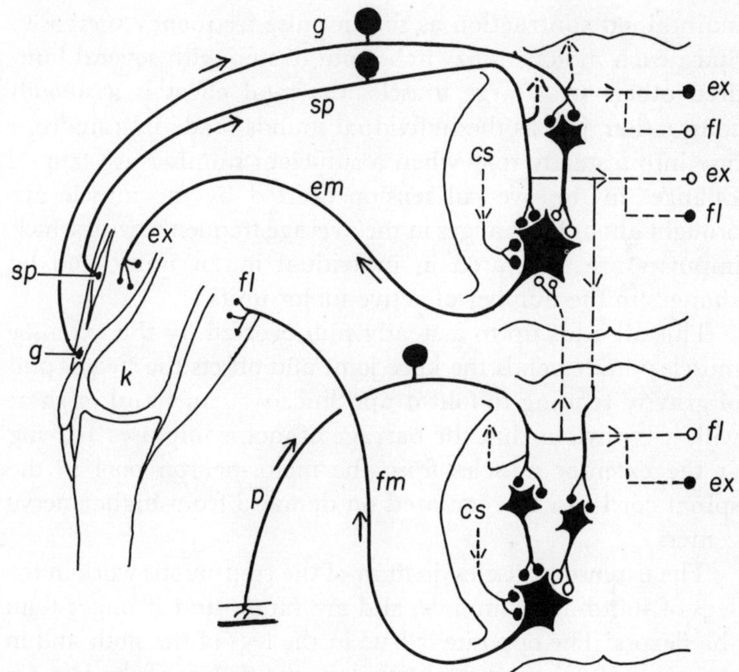

Fig. 34. Functional diagram of nerve connections to muscles and sense organs of the knee joint (*k*) shown in two segments of the vertebrate spinal cord. Excitatory synapses, *solid circles;* inhibitory synapses, *open circles.* The stretch-reflex mechanism operates through the spindle and its sensory fiber *sp,* extensor motor neuron *em,* and extensor muscles of the thigh *ex.* The tendon reflex operates through the Golgi tendon organ and sensory fiber *g* and the inhibitory interneuron. The flexion reflex operates through the pain ending in the skin and its sensory fiber *p,* the flexor motor neuron *fm,* and the flexor muscles *fl.* The dashed lines suggest some of the connections made with other regions. All three sensory systems form tracts ascending to the brain (*up-pointing arrows*); they act reciprocally with motor units of the opposite leg (*horizontal dashed lines*), and reciprocally with the ipselateral extensor and flexor motor neurons (*vertical dashed lines*). Motor neurons are also excited by corticospinal tracts *cs* from the brain. (Modified after Eccles.[10])

making up the muscle. A series of impulses originating in the motor neuron of this *motor unit* causes a series of twitches in its muscle fibers. These twitches fuse to some extent into a maintained contraction as the impulse frequency increases. Since each motor unit twitches out of step with several hundred others in a large muscle, the total effect is a smooth contraction, just as the individual sounds made by raindrops fuse into a steady roar when a sufficient number are falling. Changes in the over-all tension exerted by the muscle are brought about by changes in the average frequency with which impulses are generated in individual motor units, and by changes in the number of active motor units.

This all adds up to a steady pull exerted by the extensor muscles that extends the knee joint and offsets the steady pull of gravity tending to fold it up. Since we can stand or sit at will, it is obvious that the barrage of motor impulses arriving at the extensor muscles from the motor-neuron pool in the spinal cord can be regulated on demand from higher nerve centers.

The extensor muscles do most of the continuous work in the legs of standing mammals, and are larger and stronger than the flexors. The opposite is true in the legs of the sloth and in the arms of primates. Nevertheless, the flexors of the legs are in a state of continuous readiness and slight contraction even during standing. They have similar motor-neuron pools in the spinal cord that are excited in an exactly reciprocal fashion to those of the extensors. Any input of nerve impulses that excites the extensor neurons invariably inhibits the flexor neurons, and vice versa (Fig. 34).

This antigravity mechanism becomes interesting when we examine its operation in the presence of perturbations. It is regulated in many and complex ways, but only three of the simplest can be considered here. The first is the myotatic or stretch reflex. If a man is handed a weight of 50 pounds he receives it with practically no change in posture, so that the extensor muscles in each leg must automatically increase their tension from a value sufficient to offset his body weight to include also that of the additional load. This change in muscle tension is accomplished entirely without thought, and the neural events are of the simplest kind. Sense organs lying within each muscle, known as muscle spindles, contain receptors that are stimulated by changes in the muscle length. Axons from the spindle receptors in a given muscle pass back into the spinal cord and form synaptic endings on the motor neurons that determine the activity of its motor units (Fig. 34). Any tendency of the muscle to change its length from that preset by the original nerve impulses coming from the brain causes a change in the frequency of impulses discharged by the spindles; a stretching of the muscle causes more spindle impulses and a shortening causes less. In the example we have taken, the imposition of an additional load on the knee extensor muscles adjusted to offset only the body weight causes the muscles to begin to stretch. This incipient lengthening stimulates the muscle spindles to fire impulses with increased frequency into the motor-neuron pool in the spinal cord. The additional spindle impulses summate with the steady stream arriving from command centers in the brain, causing an increase in the barrage of impulses going to the motor units. This increases the tension in the muscle so as to adjust it to the increased load. When the weight is suddenly discarded, the tendency of the muscle to shorten due to the drop in tension causes a corresponding decrease in the discharge of spindle impulses and a drop in the total excitation of the motor-neuron pool.

In addition to direct additive action on the motor neurons supplying their own muscle, the spindle impulses are transmitted to other parts of the nervous system. In the present context it is important to note that they also reach the motor-neuron pool of the knee flexors (Fig. 34) through internuncial neurons, where their action is inhibitory. Therefore, an increase in extensor tension is automatically accompanied by a decrease in flexor tension.

The stretch reflex is a feedback system that maintains constant muscle length by varying muscle tension to match varying loads. It is the simplest possible reflex mechanism, and its large sensory and motor axons, interrupted by only one synapse, allow it to operate within about 20 milliseconds so as to adjust to rapidly changing loads with a minimum of lag time. However, this lag time is noticeable in the sag of the knees when a person is suddenly and unexpectedly loaded down. It is also evident in the familiar knee jerk, when the knee extensor muscles respond somewhat late, and extend the knee joint only after a tap on the tendon has caused a momentary stretch.

Overload protection of the muscle and its tendon against rupture by excessive loads is provided by mechanoreceptors in the tendon (Golgi tendon organs) that discharge impulses into the motor-neuron pool when the tendon becomes excessively stretched. Instead of summating with the impulses from the spindles and the brain, the Golgi impulses subtract from or override these, so as to inhibit the motor neurons. This interrupts the constant-length feedback, and the excessive load causes the legs to flex and collapse. The effect has been called the jackknife reflex because the sudden relaxation of the muscle that occurs when continued traction approaches the breaking point of the tendon may be likened to the sudden closing of the spring-loaded blade of a jackknife.

Another protective mechanism that may interrupt and override the state of constant length regulated by the stretch reflex is provided by the flexion reflex. If one foot encounters some sharp and noxious object, it is immediately withdrawn (flexed) and all of the body weight is thrown upon the other foot. The flexion reflex takes precedence over current behavior, like the evasive reactions of the moths and cockroach, and flexion of the leg is begun before nerve impulses have had sufficient time to arrive in the brain and intrude upon consciousness as pain. However, it is interesting to note that the flexion reflex can be partly suppressed if one consciously anticipates the pain.

The neural mechanisms of the flexion reflex (Fig. 34) are also fairly simple, as might be expected from the importance of speed in the operation of protective mechanisms of this kind. However, simplicity is a relative term. The noxious stimulus generates impulses in receptors in the skin that are transmitted over their axons to the spinal cord. Their course cannot be followed here in detail except to note that they excite internuncial neurons that in turn excite the motor neurons of the flexor muscles. These "nociceptive" impulses simultaneously inhibit the activity of the extensor motor neurons that was responsible for the standing leg, making it possible for the weaker flexor muscles to lift the leg from the ground. At the same time, the extensor tension in the other leg must be doubled if it is to bear the full weight of the body, so the extensor motor neurons of the opposite side receive excitation from the nociceptor fibers in addition to that deriving from their own stretch-reflex mechanisms.

At this point it is clear that the complexity of the situation is compounding at an alarming rate, even though we are merely considering the neural mechanisms regulating one muscle in one leg, in one segment of the spinal cord. Since our objective is to find out how neurons of the central nervous system integrate the various streams of impulses impinging on them, and not to describe the postural mechanisms of mammals, we shall turn from direct observation of these postural reactions and attempt to view them from a point within the gray matter of the spinal cord occupied by a single motor neuron controlling one of the motor units of the knee extensors.

Some of the inputs to this extensor motor neuron are diagramed in Fig. 35, which does not represent the actual topography and anatomy of the situation. For instance, each input in the diagram is in reality represented by a number of separate synaptic contacts on the neuron. Furthermore, only a fraction of the actual inputs are shown.

The axon of the motor neuron terminates in excitatory end plates on the muscle fibers of the motor unit. Its output varies only in the frequency with which impulses are generated in the cell body of the motor neuron, and determines the fre-

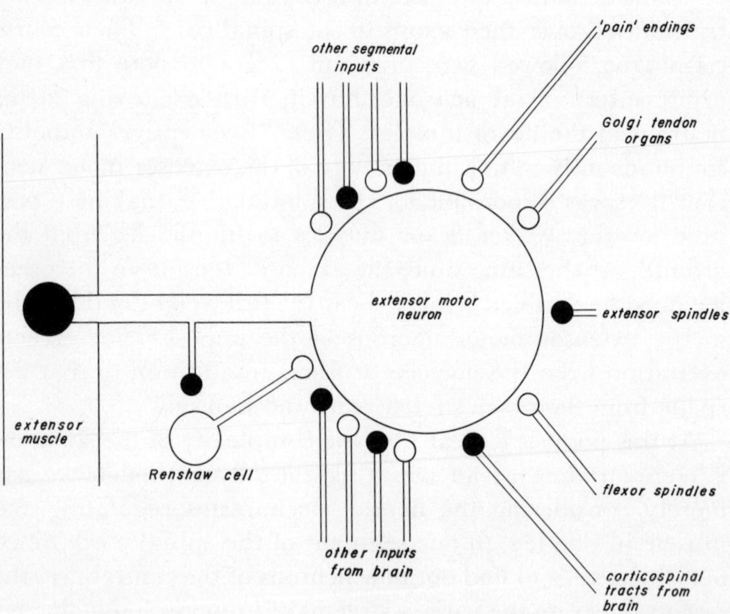

Fig. 35. Diagram of some of the synaptic inputs to one motor unit belonging to the extensor motor system of the legs. Excitatory endings, *solid circles;* inhibitory endings, *open circles.* The geometry of the cell and the relations of endings are simplified out of all proportion to the actual situation (see Fig. 36).

quency of twitching in the muscle fibers. This, added to similar unphased activity in hundreds of parallel motor units, determines the almost smooth contraction of the muscle.

The excitatory (*E*) impulses from the brain via the corticospinal tracts represent the voluntary command to stand. They do not act alone, but are supported by excitatory and inhibitory nerve activity in tracts from several other brain centers. For present purposes these have been lumped together in the diagram. Excitatory impulses from spindles in its own muscle, and to some extent from other extensor muscles in the leg, act collectively on the whole entensor motor-neuron pool to produce the stretch reflex. These are represented by one *E* input. The antagonist (flexor) muscles of the leg also have spindles, and these have some influence on the extensor motor neurons. Their action is inhibitory and is indicated by one *I* input. The inhibitory action of the Golgi tendon organs concerned in overload protection through the jack-knife reflex is indicated by an *I* input. The other protective mechanism provided by the flexion reflex is shown by an inhibitory input from the skin nerves. In addition there are several other local inputs, including those from spindles, Golgi organs, and skin endings, in the opposite leg, that act with opposite sign. These have been lumped together for the sake of brevity.

In addition to the reflex mechanisms already mentioned, the diagram includes a curious inhibitory reflex that is contained entirely within the spinal cord. The motor axon gives off short side branches that make synaptic contacts with short neurons known as Renshaw cells. The axons of the Renshaw cells return and impinge on the motor neuron. Each time an impulse passes down the motor axon it generates a rapidly declining high-frequency train of impulses in the Renshaw cell. These act in an inhibitory fashion through their junctions with the motor neuron. The negative feedback of this

Renshaw cell mechanism is thought to serve as a governor, limiting the discharge frequency of the motor neuron and preventing runaway or convulsive activity.

From the Viewpoint of the Motor Neuron. Contemplation of the situation displayed in Fig. 35 suggests that the motor neuron is a position analogous to that of an administrator. The decision is simple—in the neuron it is the frequency with which impulses are to be generated for transmission to the muscle fibers—but the decision-making involves a complex discrimination of positive and negative factors, many of which are determined by feedback, that is, by the consequences of previous decisions. It is important to note that the whole mechanism of discrimination breaks down, both in the administrator and in the neuron, if the decision is determined by a single signal acting *alone.* If any single signal were capable of determining the action, then all the others would be superfluous. Therefore, the synaptic action of these inputs must not be thought of as stimuli in the commonly accepted sense, for the arrival of an impulse at a single presynaptic terminal does not excite the motor neuron so as to generate an impulse in its axon with a one-to-one or relay type of relation. An excitatory presynaptic impulse can only facilitate (an important term in administrative argot) a postsynaptic impulse. By a similar token inhibitory impulses arriving over individual pathways cannot exert an absolute veto over postsynaptic activity; they can merely make it less probable. The output of the motor neuron is a function of the sum of facilitation and inhibition from all the inputs, including the neutrality of some of them. All the inputs are not equivalent in the amount of facilitation and inhibition they produce. Some probably have a larger number of synaptic contacts and are correspondingly more potent than others. It is probable, also, that the topographic relations of specific inputs—their contiguity or geometry in relation to one another—on the surface of the motor neuron determine greater or lesser degrees of spatial summation of their effects. If some or all of the inhibitory inputs are blocked, as through the action of strychnine, the discriminatory balance is lost and the decision-making mechanism operates like a simple relay. The administrator has become a "yes man," and the system goes into convulsion. It is tempting to liken the intimate anticonvulsive feedback provided by the Renshaw cell to the administrator's wife.

Many of these conclusions are based on discoveries made by Eccles and his associates[10] by inserting microelectrodes into single motor neurons "lying virtually unmolested in the central nervous system and being normally supplied with blood." These brilliant experiments are providing a clearer picture of the manner in which impulses arriving over some synaptic endings facilitate, while impulses in other endings inhibit, the activity of the postsynaptic neuron.

A typical motor neuron in the spinal cord of a cat (Fig. 36) has an irregularly shaped cell body or soma about 70 microns in diameter. Gradually tapered and branching dendrites may extend as far as 1 millimeter from the soma. The axon assumes a myelin sheath just beyond its origin from the soma, and it leaves the spinal cord in company with hundreds of other axons to form part of a spinal nerve. It continues without interruption to a muscle, where it branches to form motor endplates on the muscle fibers forming the motor unit.

The soma is thickly encrusted with hundreds of synaptic knobs, each the terminal enlargement of a presynaptic fiber. Each synaptic knob is about 1 micron in diameter, and separated from the soma membrane by a small but finite synaptic cleft about 200 Ångstroms wide. Each presynaptic fiber may terminate in several synaptic knobs of variable shape. The base of the axon is relatively free of synaptic knobs, although they cover the basal regions of the dendrites, becoming less dense on the dendritic branches.

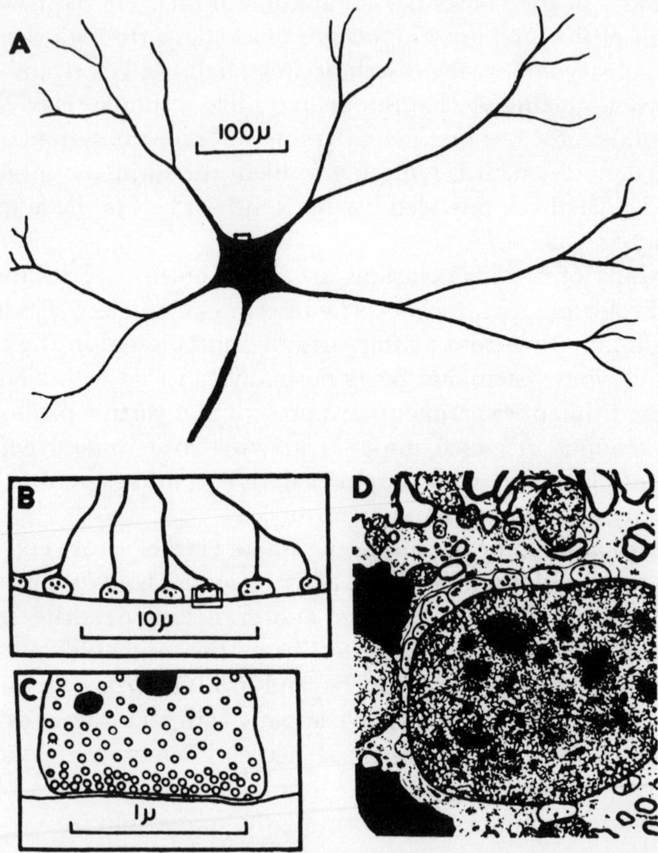

Fig. 36. Drawings of a motor neuron from the vertebrate spinal cord.[71] (*A*) The general relations of dendrites, soma (cell body), and axon. (*B*) 20-fold enlargement of the small area outlined on the soma in (*A*), showing the relation of presynaptic fibers and synaptic knobs to the cell surface. (*C*) 10-fold further enlargement of one synaptic knob outlined in (*B*), showing the cleft separating presynaptic and postsynaptic cells. (*D*) Drawing of a low-power electron micrograph showing twelve synaptic knobs in contact with a large dendrite.

The extraordinary feat of probing through several millimeters of spinal cord with a glass pipette 0.5 micron in tip diameter for an invisible target about 70 microns in diameter is hard to appreciate without a detailed knowledge of the technical problems. These details must be sought elsewhere,[10,52] since there is space here only to note a few of the pertinent results.

It was explained in Chapter 3 how the minute electric current flowing between an electrode inserted into a nerve cell or axon and an electrode placed elsewhere in the tissue but outside the cell gives a measure of the membrane potential of the cell. The inserted micropipette shows that the membrane potential of the soma of a motor neuron is about −70 millivolts (inside negative). The genesis of an impulse is accompanied by an action potential of about 100-120 millivolts lasting about 1 millisecond (Fig. 37). As in axons, the membrane potential drops rapidly to zero and then reverses in sign, momentarily reaching a value of +30 to +40 millivolts (inside positive) at the peak of the action potential. This is followed by almost complete return to the resting value of about −70 millivolts in approximately 0.5 millisecond.

A series of ingenious experiments with a double-barreled micropipette enabled Eccles to pass electric current, or to inject sodium, potassium, or chloride ions through one barrel, and to measure their effects on the membrane potential via the other. These experiments showed that the ion gradients and fluxes responsible for the membrane potential and spike potential in the soma of a motor neuron are very similar to those already known to occur in axons (Fig. 3).

In the resting membrane of the soma sodium ions are actively ejected as fast as they enter, and the resting membrane potential of −70 millivolts is mainly an expression of the tendency of potassium to diffuse outward and of chloride to diffuse inward. However—and this is important—the resting membrane potential is not entirely determined by the physical diffusion tendencies of these ions. A calculation, based on the actual concentrations of potassium and chloride present, shows that the membrane potential should be about −80 millivolts (Fig. 37) if this were the case. Eccles concludes that, in addition to actively ejecting or pumping out sodium ions as fast as they enter, the soma membrane is also actively accumulating or pumping in potassium ions. In other words, if the mechanism pumping in potassium were to stop, the net increase in the outward movement of positively charged potassium ions would increase the membrane potential from its normal value of about −70 millivolts to a value of about −80 millivolts. The outside would become more positive. This can be compared with the opposite effect to be expected if the mechanism pumping out sodium were to stop functioning. This would produce a large net influx of positively charged sodium ions so as to decrease the membrane potential from its resting value of −70 millivolts to about +40 millivolts, with the inside of the membrane positive to the outside. This latter change is equivalent to the occurrence of the spike potential (Chapter 3), and its direction is associated with excitation. The much smaller increase in membrane potential due to a decrease in the accumulation of potassium is associated with inhibition. The nature of the so-called pumping mechanism that accumulates potassium and ejects sodium is not at all clear, but it might be pointed out that a leaky ship can sink either because the bilge pumps stop working or because the sudden appearance of a much larger leak counteracts their effectiveness.

From this it follows that the resting potential (V_m, Fig. 37) of the soma membrane is not indicative of a state of rest, but rather of a dynamic equilibrium between (*a*) the combined diffusion pressures of Na^+ inward, K^+ outward, (*b*) the selective permeabilities of the membrane to these and to other ions, and (*c*) the activity of the membrane in ejecting sodium ions

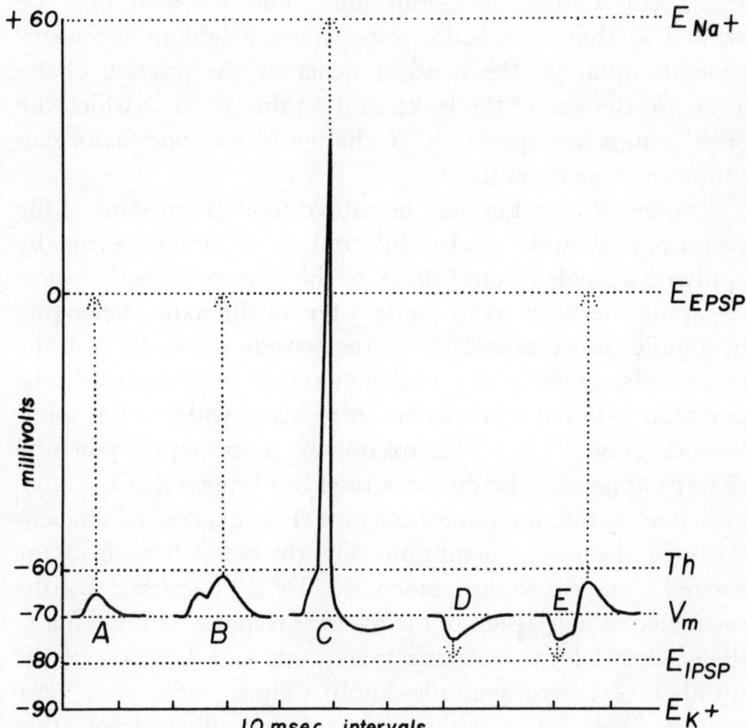

Fig. 37. Electrical responses recorded intracellularly from a motor neuron following the arrival of presynaptic volleys. (*A*) Subthreshold EPSP from a single excitatory volley; (*B*) summed EPSPs from two volleys separated in time; (*C*) threshold EPSP that generates a spike potential; (*D*) IPSP from an inhibitory volley; (*E*) IPSP summed with a later-arriving EPSP. Horizontal dotted lines, potentials the cell membrane would theoretically reach if the ions indicated diffused without restriction. Vertical dotted arrows, trends of the potentials indicated. E_{Na^+}, equilibrium potential for sodium ions; E_{K^+}, for potassium ions; E_{EPSP}, for free movement of all ions; E_{IPSP}, for potassium and chloride ions; V_m, resting-membrane potential of neuron; *Th,* membrane potential at threshold, that is, when the spike occurs. (Modified from Ruch *et al.*[52])

and accumulating potassium ions. The situation may be likened to that in a leaky ship, whose height in the water depends upon (*a*) the head of water at the position of the leaks, (*b*) the size of the leaks, and (*c*) the speed at which the bilge pumps are operating. A change in any one factor can cause a change in level.

A volley of impulses may be caused to arrive at some of the excitatory synaptic knobs clustered on a motor neuron by applying a single electric shock to the sensory (dorsal) root of the spinal nerve so as to excite a few of the axons belonging to spindles in its muscle. A microelectrode whose tip is in the soma registers this event as a momentary drop in membrane potential followed by exponential recovery within a few milliseconds (Fig. 37*A*). This excitatory postsynaptic potential (EPSP) appears to be due to a brief but intense leakage of all ions, mainly sodium, potassium, and chloride, created momentarily in the soma membrane directly beneath each of the excited synaptic knobs. Since the EPSP registered by the microelectrode is a sign of the average response of the soma to these several local leaks, its magnitude is a function of the number of active synaptic knobs. Thus, activity in more spindle fibers, or in additional excitatory fibers from other sources, produces a larger EPSP, providing a means for spatial summation. Similarly, if two synaptic volleys of impulses are timed so that the EPSP caused by the second overlaps what is left of the EPSP caused by the first, then the total EPSP formed by the sum of the two (or more) provides a basis for temporal summation (Fig. 37*B*).

A spike potential is generated when the membrane potential of the soma is lowered to a critical value (threshold) by the EPSP generated by single or combined volleys of excitatory impulses. The spike potential appears to be caused by a sudden massive influx of sodium ions, as in the spike potential of an axon (Chapter 3). The spike causes the membrane potential to overshoot and reverse in sign, so that it approaches the equilibrium point for sodium ions before the sodium influx is checked and the resting membrane potential is restored (Fig. 37*C*). The soma spike invades the axon, and is transmitted to the muscle fibers.

An electric shock applied to a nerve known to contain fibers having inhibitory connections with a motor neuron, for example, Golgi-organ fibers from the tendon of its muscle or spindle fibers from its antagonist muscle, causes the soma to develop an inhibitory postsynaptic potential (IPSP). During the IPSP the membrane potential of the soma increases, approaching a value of −80 millivolts (Fig. 37*D*). It appears that the arrival of impulses at inhibitory synaptic knobs causes local leaks in the soma membrane adjacent to each synaptic knob, but in this case the leakage is restricted specifically to potassium and chloride ions. The mechanism for sodium exclusion remains unaffected. Since the potassium pump maintains the membrane potential at −70 millivolts in the face of the tendency of K^+ to diffuse outward and increase it to a value of about −80 millivolts, the brief additional leakage of potassium outward under the inhibitory synaptic knobs causes a momentary swing of the membrane in the direction of −80 millivolts.

Temporal and spatial summation of IPSPs can be demonstrated by stimulating the same or separate inhibitory pathways twice in quick succession. Nevertheless, it is clear from Fig. 37*D* that an IPSP can never normally generate an impulse, since it causes the membrane potential to move away from its threshold. The effect of an IPSP becomes manifest only when it summates with an EPSP (Fig. 37*E*). In this example, an IPSP has been generated in a similar manner to that shown in (*D*), and 1 to 2 milliseconds later an EPSP has been generated by stimulating excitatory presynaptic fibers. The EPSP is the same size as that in (*C*), which depressed the membrane potential sufficiently to elicit a spike potential from the soma.

However, when it is superimposed on the increased membrane potential produced by the preceding IPSP, the EPSP fails to bring the membrane potential to the threshold level, and the inhibitory volley has been effective.

EPSPs and IPSPs generated by the artificial method of applying single electric shocks to groups of excitatory and inhibitory nerve fibers gives an exciting but still most incomplete glimpse of the natural processes of integration. Natural excitation of sensory endings rarely produces the more-or-less synchronous volleys of presynaptic impulses generated by electric stimulation. The greater part of the natural presynaptic input converging on a motor neuron probably consists of unsynchronized trains of excitatory and inhibitory impulses that separately wax and wane, sometimes gradually and sometimes abruptly. This means that, instead of the discrete EPSPs and IPSPs generated by brief electrical stimuli, these changes must cause continuous, gradual or, occasionally, abrupt, fluctuations in the membrane potential around its resting value of about −70 millivolts. When one of these fluctuations carries the membrane potential to the critical or threshold level of −60 to −65 millivolts a spike potential erupts and is transmitted to the axon. This event nullifies all previous integration through the combined action of the refractory period that accompanies the spike and the longer-acting inhibition backfired on the soma from the stimulated Renshaw cells. As excitability of the soma returns, this complex summation in time and space of excitatory and inhibitory influences once more determines the timing of the next impulse.

Contemplation of the picture painted by these intracellular studies suggests that the interplay of presynaptic events impinging on a single unit of the nervous system is immensely complex in space as well as in time. The actual shape of a spinal motor neuron (Fig. 36) departs widely from the simple sphere diagramed in Fig. 35. Thicker portions of the dendrites as well as the highly irregular soma are encrusted with hundreds or thousands of presynaptic terminals. It is impossible to visualize the kinds of spatial interactions between EPSPs and IPSPs that must take place on this complex topography. Farther out the dendrites branch into elaborate and fine arborizations, particularly in certain neurons found in the brain. The role of these arborizations in influencing the spike discharge of the neuron is completely unknown. Lettvin has recently pointed out (personal communication) that if this spatial complexity is compounded with the temporal complexity of arriving nerve volleys the result not only staggers the imagination but makes nonsense out of any current attempts at quantitative description of the natural situation surrounding a single nerve unit. Intracellular studies of single units are valuable because they have made us more soberly aware of the actual situation, but they will need the assistance of all the traditional and varied attempts to relate nerve activity to behavior, as well as inestimable novelty in the research of tomorrow, before we can even faintly comprehend the operation of the behaving system into which the nerve units are woven. It is strangely exhilarating to realize that the anticlimax of understanding is still a long way off.

Mechanism of Synaptic Action. At the present time, the most plausible explanation of the separate membrane changes caused by excitatory and inhibitory inputs is that the arrival of impulses at their respective synaptic knobs triggers the release of specific excitatory and inhibitory mediator substances into the subsynaptic clefts. Over 30 years ago, Dr. Otto Loewi demonstrated that a chemical indentical to acetylcholine is released by the vagal (inhibitory) fibers to the heart, and that a substance allied to adrenalin is released by the cardiac accelerator fibers. Since then acetylcholine has been identified as the excitatory transmitter released at certain autonomic endings, and at the end-plates of motor axons on the striated-muscle fibers or vertebrates.

These findings are accepted fairly widely, but it has been much harder to determine whether chemical mediation plays a major part in neuronal interaction within the central nervous system. The direct demonstration of separate excitatory and inhibitory substances requires that they be chemically isolated and identified, and that their actions be separately tested. The extreme difficulty of this chemical separation becomes more apparent when it is realized that during normal bombardment of a motor neuron by presynaptic impulses two and probably more separate chemical entities, each with effective lives of 1 millisecond or less, are being ejected in brief squirts over distances of about 200 Ångstroms at a number of sites each about 1 micron in diameter and scattered in crazy-quilt fashion over the surface of the 70-micron soma.

Indirect evidence is strong and copious.[10] Vertebrate motor neurons are known to excite the muscle fibers of their motor units by releasing acetylcholine at the axon terminations, and it is highly probable that the Renshaw cells are excited in the same manner. The convulsant substances strychnine and tetanus toxin appear to owe their action to a highly selective blockade of inhibitory processes in the vertebrate spinal cord, suggesting that they interfere with the action of an unidentified inhibitory mediator. There is considerable indirect evidence that acetylcholine is not the only excitatory mediator, and neuropharmacology has produced a host of candidate substances for both excitation and inhibition. The growing multitude of neuropharmacological agents, such as convulsants, anticonvulsants, hallucinogens, tranquilizers, and so forth, probably owe their specific and different actions to interference with specific operations within the central nervous system, and the most logical spot for this interference would be at the synaptic knobs concerned in some specific integrative mechanism.

A Spectrum of Interactions. At this point it is necessary to withdraw from closer scrutiny of these mechanisms lest we become embroiled in controversies regarding the relative importance of electrical and chemical factors in nerve activity. However, it seems that the lines of battle drawn up between the classical exponents of the "soup" and "sparks" theories are fast disappearing. Communication and discrimination in the nervous system appear to be based on a spectrum of mechanisms ranging from the closely coupled "repeater" interaction through ion currents flowing between adjacent parts of the axon membrane, through chemical actions with a time course of the order of 1 millisecond and a range of 0.02 micron at the juxtaposition of cell membranes under a synaptic knob, to the traditional hormones acting with a time course measured in days or weeks and a range having the dimensions of the body. These mechanisms alternate at different nodes in the network of communication.

The apparent distinctiveness of each of these mechanisms may be an illusion created by the great differences in their dimensions, and they may ultimately prove to be segments arbitrarily selected from a continuous spectrum of chemical interaction. Already, the range of this spectrum has been extended at one end by our increasing knowledge of pheromones. These are "substances that are secreted by an animal to the outside and cause a specific reaction in a receiving animal of the same species."[20] Chemical releasers have been known for a long time, for example, as sex attractants in connection with the courtship of moths and many other animals. More recently they have been shown to be used for foraging bees in trail-marking,[22] while ants appear to possess a "vocabulary" of different pheromones indicating food trails, danger, nest mate, queen, and even death.[70]

At the other end of the spectrum the literature contains a number of hints of local interaction between individual cells. Wigglesworth has recently[68] discussed the importance of cell-to-cell contact in the growth and regeneration of insect epidermis, and envisions a direct chemical interchange between adjacent cells. This sort of mechanism is similar to that postulated for the well-known "organizers" in embryonic growth, and may well represent the primordial system of cellular coordination. At another point, it seems possible that the gap between the mechanisms of impulse conduction in axons and the synaptic processes may be bridged by simple "repeater" transmission between neurons, in which ion flow and chemical action combine to form the trigger mechanism. At still another level there are several suggestions of mechanisms operating at a more diffuse level than synaptic transmission, but more locally than circulation-borne hormones. Natural groups of neurons forming a "pool," ganglion, or center within the nervous system sometimes show evidence of being coupled in their activity without the interchange of propagated impulses. There are suggestions that this coupling is accomplished by local electrical and chemical fields generated by the group activity, and even by direct low-resistance pathways connecting individual neurons.[64]

In conclusion, it must be observed that what might appear to have been a penetration in depth as far as the topic of this book is concerned has in reality penetrated no deeper than a single motor neuron in the lowliest part of the vertebrate central nervous system. Its mechanisms have been examined only in relation to the simplest of reflexes, and it has been considered in isolation from the other members of its motor neuron pool. Yet it is about as far as we can go in this direction in attempting to link neurophysiological events with behavior.

9. *Endogenous Activity of Neurons*

The concentration of research talent on the mammalian spinal cord that produced the elegant and penetrating discoveries outlined in the preceding chapter must be laid in part to the inevitable and natural concern of man with his own species. I suspect that our basic humanism—that feeling of greatest subjective rapport with those animals most closely related to us—has as much to do with this as the intrinsic scientific value of these experiments and their practical importance to medical science. However, comparison has been the underlying principle of zoology since, and even before, the time of Darwin. In his study of life the zoologist has never been able to derive a sense of false security from ultimate standards or constants, such as the physicists once had in electrons, protons, and the velocity of light. Life cannot yet be defined, so that a comparison of species morphology, physiology, behavior, biochemistry, genetics, and neural mechanisms has always been the most intellectually rewarding approach in zoology. This provides some degree of counterbalance to our entirely natural egocentrism.

It is necessary to turn from these cozy generalizations to the sober fact that our knowledge of the activity of single neurons in insect ganglia is rudimentary compared with what is known about motor neurons in the mammalian spinal cord. It is my opinion that if as much were known about the neural mechanisms in an insect ganglion, the advance obtained in terms of understanding the basis of animal behavior would be comparatively greater, since an insect behaves by employing fewer neurons than does a vertebrate.

Insect Neurons. The only reason for this lack of information is that it has not been sought by a sufficient number of investigators. When a full-fledged microelectrode study of an insect ganglion is eventually undertaken, it must not be based on the assumption that the situation will be similar to that in the mammalian spinal cord. To begin with, the anatomic relations of neurons in invertebrate ganglia are very different. Insect motor and internuncial neurons are monopolar, the cell bodies being on a side branch from the receiving and conducting parts (Fig. 2). The cell body does not appear to be a site for the synaptic knobs and thus concerned directly in the integrating mechanism, but instead it lies together with others in the external layer or cortex of the ganglion. Each cell body sends a process into the central part of the ganglion, contributing to a dense felt of fibers known as the neuropile (Fig. 38). Details of the synaptic contacts are unknown, but sensory and internuncial neurons appear to branch into a series of fine terminals in the neuropile, where they make close contact with fine dendritic branches of the postsynaptic fibers. Terminal regions of pre- and postsynaptic fibers have such small diameters that it will be difficult to achieve intracellular penetration with the microelectrodes at present available. However, there would be considerable value in a systematic extracellular exploration of the neuropile in relation to specific segmental reflexes. The cell bodies of the internuncial and motor neurons in the cortical regions of the ganglion are large, and should be readily accessible to intracellular recording. However, no one has yet attempted to find out whether these cell bodies are in-

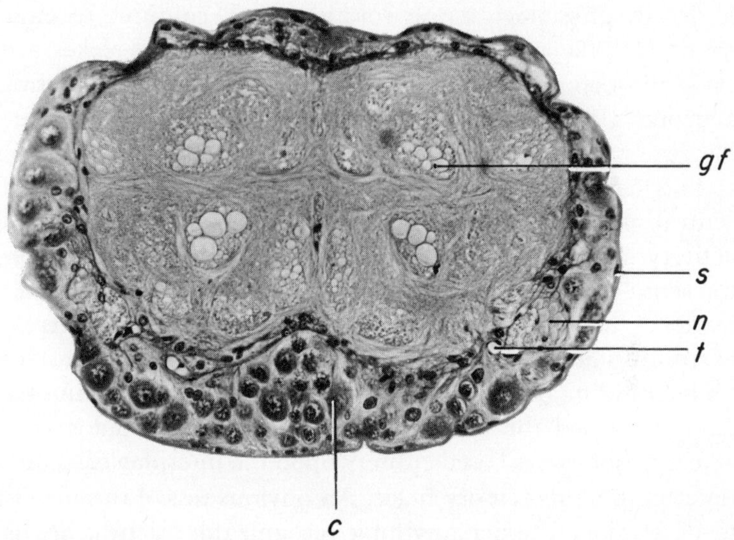

Fig. 38. Cross section of the abdominal ganglion of the cockroach *Periplaneta americana*. *c,* cell bodies of unipolar motor and internuncial neurons. A process from the cell under the guide line can be seen extending toward the neuropile. *g,* bundles of giant fibers (compare Fig. 2) passing through the ganglion. Many other tracts and commissures, as well as granular-appearing regions probably occupied by synaptic contacts, make up the central neuropile. *n,* nerve about to enter the ganglion; *s,* sheath or neural lamella surrounding the ganglion; *t,* trachea.

vaded by impulses generated in the neuropile, or whether an electrode implanted in them might perhaps provide a point for monitoring synaptic events occurring deeper in the ganglion (see Chapter 11).

In contrast to this backward state of affairs, invertebrates provided the first definitive evidence of endogenous or spontaneous activity in individual neurons of the central nervous system, and they have recently provided some hints as to its significance in animal behavior. In 1930 Adrian[1] noted that if the ventral nerve cords of a caterpillar and a beetle are completely dissected out of the insects' bodies and placed on electrodes in physiological salt solution, they continue to emit trains of spike potentials for hours or days. The spikes are evidently being generated recurrently in a number of central neurons, mostly firing asynchronously with one another, but sometimes combining in regular bursts.

This continuing nerve activity in isolated portions of the central nervous system was originally termed spontaneous activity because it is not generated by stimuli impinging on the sense organs. Until fairly recently there was a widespread tendency to consider elementary behavior as a matter merely of stimulus and response. This idea was based on an incorrect understanding of conditioned and spinal reflex mechanisms, for it regarded the animal as a highly complex automaton whose actions were based entirely upon the interplay of stimuli impinging on its sensory fields. An obvious test of this idea is to determine whether anything recognizable as behavior is displayed by an animal when placed under circumstances of complete sensory deprivation, a condition almost impossible to achieve either by surgery or by environmental isolation in outer space.[37]

The undeniable demonstration of continuing activity of neurons in the isolated central nervous system found no place as a factor in behavior under the stimulus-response concept, and it was dismissed by some as being analogous to the "noise" in a radio receiver or amplifier. The term "spontaneous activity" was taken by some to imply activity occurring "in vacuo," or without any cause. However, the experiments with isolated nervous systems of invertebrates have demonstrated merely that spikes continue to be generated by central neurons following elimination of all afferent impulses from the sensory receptors, and it has also been shown by many investigators[37] that the level of this activity is highly sensitive to internal changes such as the chemical composition[35] of the solution bathing the central nervous system. The fact that this activity is independent of changes ordinarily originating outside the animal, but very dependent upon the internal chemical state, suggests that "endogenous activity" is a more suitable term.

Before looking more closely at the relation between endogenous activity in neurons and that in the behaving animals of which they are a part, we shall find it helpful to examine the general relation between poised systems containing potential energy, such as axons, and endogenously active systems.

Models. There are several types of physical system that might provide insight into the mechanism of endogenous activity. Two that are important are regeneration and relaxation oscillation. Regeneration is easily understood by contemplating the escapement mechanism of an ordinary clock, having either a weight or a spring as its store of potential energy (Fig. 39*A*). When this energy store is empty, the balance wheel or pendulum can be made to oscillate with its natural period, but

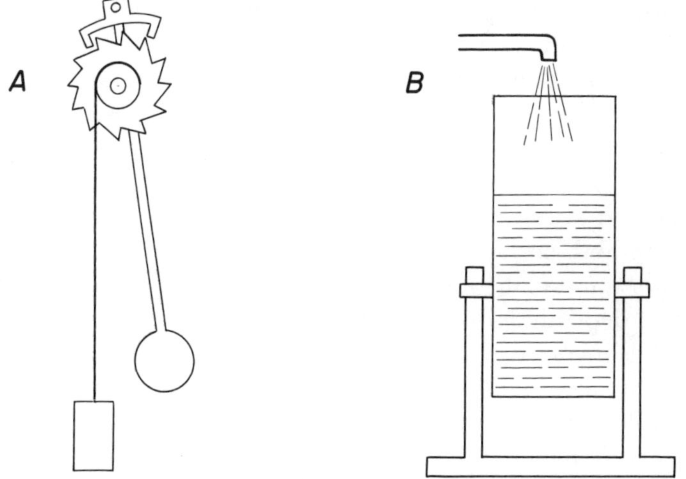

Fig. 39. Mechanical analogs for (*A*) regenerative and (*B*) relaxation oscillators.

the oscillation is damped and dies away because of friction. When the clock has been wound, the tension of the weight or spring is converted into kinetic energy by the escapement in such a manner that this energy is always *added* to that in the pendulum by giving it a small impulse in the right direction during each swing. The frequency of oscillation (the accuracy of the clock) is relatively independent of the magnitude of the force exerted by the weight or spring provided that this is adequate, but is directly dependent upon the moment of inertia of the pendulum or balance wheel. The system may be said to depend upon the application of a small amount of positive feedback to the oscillating system. Many types of electronic and mechanical oscillator depend upon the regenerative principle, and their outputs, like the movement of the pendulum, are commonly, although not necessarily, sinusoidal.

Relaxation oscillators, or fly-back devices, also convert a steady supply of energy into oscillations, but their principles and output differ from those of regenerative oscillators. A simple model is the swinging bucket shown in Fig. 39*B*. Its fulcrum is below its mid-point, but when empty it hangs in an upright position because of the weight of its bottom. As it fills steadily with water there is no detectable change until it becomes top-heavy; then it tips, empties completely, and returns to an upright position. This relaxation oscillation is repeated at intervals that depend upon the rate of inflowing water and on the capacity of the bucket. The oscillation is nonsinusoidal, and may be saw-toothed or pulselike in form, depending upon the shape of the bucket. Positive feedback is involved, since when the bucket fills to the point of instability any slight deviation promotes a greater deviation and complete tipping. It is another model of the avalanche phenomenon pictured in connection with spike generation in the axon (Chapter 3). A leak in the bucket will lengthen the intervals between oscillations, and if it is large enough that outflow equals inflow before the water level reaches the unstable point the leak can prevent further oscillation. The bucket then remains stationary, although a slight impulse from outside can precipitate one complete oscillation. There are many types of mechanical and electronic relaxation oscillator, including the familiar flashers made from neon-filled glow-discharge lamps.

Endogenous Mechanisms in Neurons. It seems probable that both regeneration and relaxation oscillation are involved in the endogenous activity of neurons. Sinusoidal nonpropagated oscillations of membrane potential have been observed to occur naturally, and sequences of undamped oscillations terminating in a propagated spike can be induced by chemical means, particularly a reduction in the calcium content of the surrounding medium. Even isolated axons, normally inactive unless invaded by impulses from their cell bodies, can be made to discharge endogenous impulses if the calcium content of their surroundings is reduced.[6] This effect is perfectly reversible, inactivity being restored by immersion in saline solution containing calcium chloride. The sequences of endogenous spikes recorded from the isolated nerve cords of insects and crustacea may continue for hours or days when the cords are immersed in the animal's blood or in a solution of similar salt composition. However, the number of active neurons and the frequency of spikes can be systematically and markedly altered by changes in the chemical composition of the medium.[35]

At the end of Chapter 8 it was pointed out that it is incorrect to think of individual presynaptic impulses as stimulating or inhibiting a motor neuron. They merely add to or subtract from its state of poised instability. Its instantaneous state is determined by the sum of these inputs plus the action of its chemical environment and its previous history of activity. This state determines the frequency with which it generates impulses. A glance at Fig. 37 shows that any factor reducing the membrane potential (V_m) to the threshold value would cause the neuron to discharge a series of spikes limited in frequency by the refractory period, and by the afterpositivity due to inhibitory feedback from the Renshaw cells. Indeed, the function of the latter may well be to curb such runaway activity. There are many chemical changes, other than those produced by incident EPSPs, that could bring about a constant reduction in membrane potential. Regarded from this viewpoint, the distinction between synaptically evoked and endogenous activity becomes much less sharp. Indeed, to consider means whereby endogenous factors are prevented from preempting the output of a neuron, so as to leave some control to sensory input, seems as important as it is to determine the basis of endogenous activity.

In the preceding paragraphs the endogenous units were presumed to be single neurons in which the mechanisms for regenerative interaction or relaxation oscillation resided in the cell membrane and its immediate surroundings. While this appears to be the case with much of the endogenous nerve activity in invertebrate nervous systems, there are other examples of endogenous activity to be found within the central nervous system where the active system is composed of closed chains or loops containing several neurons in synaptic contact. For instance, a loop similar to that formed by the motor neuron and Renshaw cell, with two excitatory synapses instead of one excitatory and one inhibitory synapse, could "trap" an impulse so that it circulated indefinitely. The loop would have to be long enough (or conduction sufficiently delayed) that each neuron could recover its excitability before it was reinvaded by the impulse. Loops with this sort of positive feedback have been described in the mammalian central nervous system, and because of their ability to trap circulating impulses they have been evoked as the mechanism responsible for short-term memory. It seems to me that this would be a very uneconomical way of storing information.

Another type of oscillation that is self-sustaining, although it is not confined to the central nervous system, is possible during abnormal operation of any one of the many negative-feedback systems that regulate bodily activities. For instance, the stretch reflex mentioned in Chapter 8 operates so as to keep the extensor muscles concerned in posture at constant length by making rapid adjustments in muscle tension to match changes in external load. The large-diameter axons and the monosynaptic nature of the stretch-reflex mechanism reduce delay in the feedback from the spindles to a minimum, but if sufficient lag were introduced into the feedback, or if the amount of feedback was excessive, it is probable that instead of maintaining a constant length the muscle would go into an uncontrollable series of oscillations.

The reason for this is apparent if we resort to a more familiar analogy. An experienced automobile driver steers an (almost) straight course along the road by applying continuous and minor corrections via the steering wheel. He constitutes the feedback that corrects the course for perturbations (irregularities in the road, gusts of wind) as they are encountered by immediately applying just sufficient compensation to the wheel. A learner, unfamiliar with the performance of the car, does not notice a deviation toward the ditch until the car is almost off the road. Too late, he turns the wheel too far so that the car swerves to the other side, and the oscillation has begun. Although this type of oscillation may be triggered by some minor external factor, it is maintained by the internal defects of the system, and must be classed as endogenous.

Endogenous Activity in Receptors. In addition to neurons within the central nervous system, a number of peripherally placed sensory receptor cells have been shown to be endogenously active. This may seem to be a paradoxical statement in terms of the definition of endogenous activity given above, but it implies merely that the receptor cells continue to generate impulses when completely shielded from their normal sources of stimulation. Sometimes such shielding is easy to achieve, as in the case of photoreceptors; often it is quite difficult, as in the case of mechanoreceptors.

Invertebrates have supplied many examples of endogenously active receptor cells, but, following the practice of this book, examples will be selected from among the insects. One case has already been discussed (Chapter 4) in the *A* and *B* receptor cells of the moth's ear. During the absence of ultrasonic stimulation the *A* cells discharge impulses in the tympanic-nerve fibers in what appears to be a random pattern. During silence this random pattern continues indefinitely. An ultrasonic tone causes a regular discharge at higher frequency. Presumbly temporal summation of the *A* impulses at synapses within the central nervous system of the moth determines the significant spike frequency in terms of ultimate behavior. Exposure of the ear to a continuous tone causes the *A* cells to adapt, as indicated by their declining discharge frequency (Fig. 9). This loss in sensitivity produced by continued stimulation persists even after the tone has ceased, for the endogenous discharge is absent for some time following the "off," and returns gradually. Tests with short ultrasonic pulses applied at various times during this return of endogenous activity show that it coincides with the disappearance of adaptation, or with what is more conveniently called the process of disadaptation.

It is possible that the continued and regular spike discharge in the *B* cell (Fig. 12) is also a form of endogenous activity, but, since the stimuli to which the *B* cell normally reacts have not been identified, this question cannot now be answered.

An interesting and more complex case of endogenous activity in a sensory mechanism has recently been brought to light by Ruck[53] in the ocelli of dragonflies and cockroaches. The ocelli are relatively simple photoreceptors of great sensitivity found on the mid-dorsal region of the head in many insects. They are entirely separate from the great compound eyes used in visual orientation.

An ocellus is shown diagrammatically in Fig. 40. Several hundred photoreceptor cells are arrayed beneath the highly convex cornea. Short axons converging from each receptor cell form synaptic contacts with four or five large ocellar nerves that connect the ocellus with the brain.

Analysis of the polyphasic potential changes detected by electrodes placed at various levels within the ocellus and on the ocellar nerve led Ruck to conclude that illumination of the ocellus causes stimulation of the receptor cells and the generation of impulses in their short axons. The arrival of these impulses at the receptor-cell terminations brings about inhibition

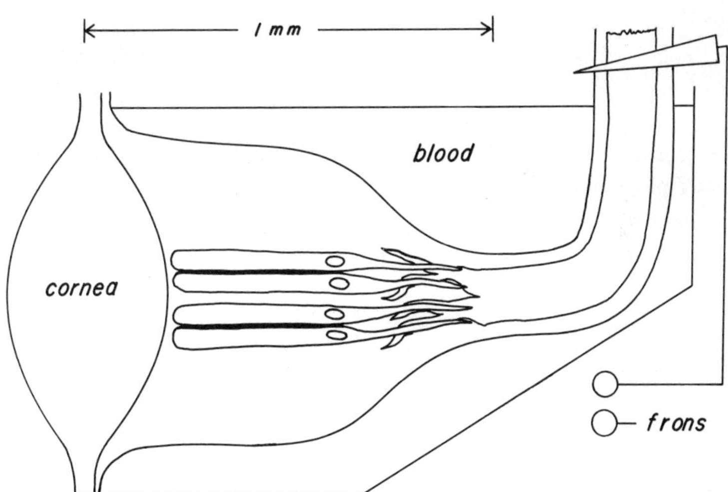

Fig. 40. Diagram of the lateral ocellus of the dragonfly showing the method of recording the potentials of Fig. 32. Synapses made by two pairs of photoreceptor cells with a large ocellar-nerve fiber are shown. The nerve is severed, and a clamp serves as electrode and to lift the nerve stump out of the blood bathing the ocellus. The indifferent electrode makes contact with the blood beneath the frons. (Modified from Ruck.[53])

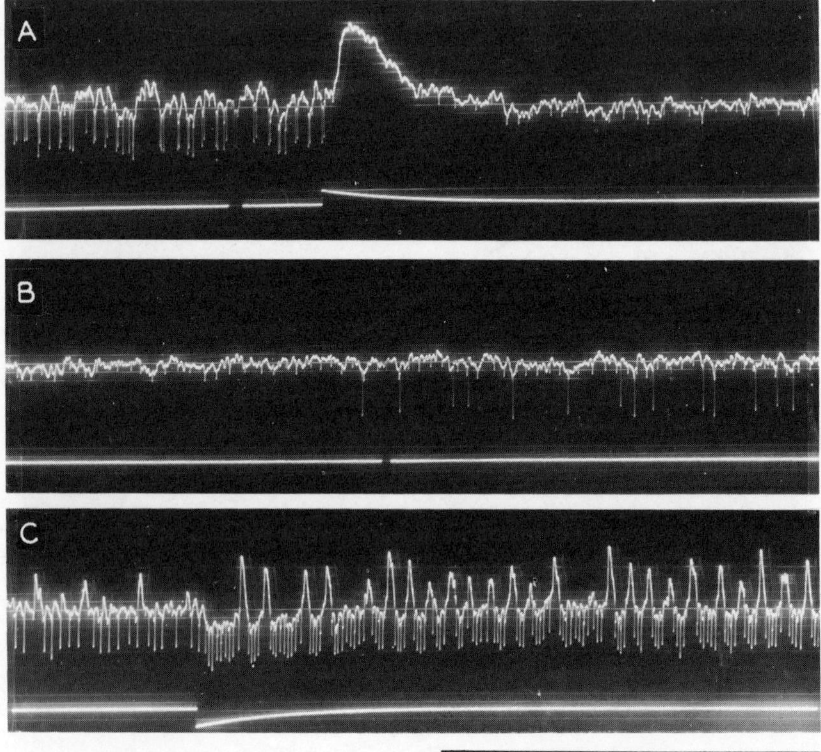

Fig. 41. Activity from the lateral ocellar nerve of the dragonfly *Libellula vibrans.*[53] (*A*) Dark-adapted state with irregular endogenous activity of photoreceptor cells (*small upward deflections*) and ocellar-nerve fibers (*spikelike downward deflections*). The ocellus is illuminated at the upward deflection of the signal (*lower line*). (*B*) Continuation of the record. (*C*) The same preparation, slightly deteriorated; the light is turned off at the signal; regular endogenous volleying of photoreceptor cells (*upward deflections*) coincides with inhibition of endogenous activity in ocellar-nerve fibers.

of an endogenous discharge (Fig. 41, downward spikes) that originates in the distal terminals of the ocellar-nerve fibers and is transmitted to the brain of the dragonfly. Thus, the impulse discharge in the ocellar-nerve fibers diminishes or ceases entirely when the ocellus is illuminated (Fig. 41*A*). In continued illumination the receptor cells partially adapt, allowing the ocellar-nerve fibers partially to escape from inhibition, so that a few ocellar-nerve spikes are generated (Fig. 41*B*). In some cases random discharges by the photoreceptor cells occur even during illumination. In Fig. 41*C* these are evident as upward deflections, during which the ocellar-nerve spikes are suppressed. With the onset of darkness there is an increase in random discharges from the photoreceptor cells as well as from the ocellar-nerve fibers upon which they act. The result is a high-frequency discharge of spikes in the ocellar nerve, broken at frequent intervals by inhibition from the photoreceptors. In record (*C*) the preparation had deteriorated somewhat, and it is possible that some of the several hundred photoreceptor cells were no longer responding to illumination, but it was chosen because it shows in exaggerated form the negative interaction between these two endogenously active systems.

Unfortunately, next to nothing is known about the role of the ocellus in insect behavior. Its sensitivity to light is one or two orders of magnitude greater than that of the compound eyes, but it seems to be unfitted to transmit information on form discrimination. If, as appears, the ocellus is playing the part merely of a highly sensitive light detector, it is hard to understand why such a complex method of signaling, employing two levels of endogenous activity with excitatory and inhibitory links, is necessary in order to convey information on light intensity to the brain of the insect. Furthermore, there is no obvious correlation between the reversal of sign—more light, fewer ocellar-nerve impulses—and the behavior or habits of the species. Dragonflies are generally diurnal animals and become inactive at night, but the ocelli of predominantly nocturnal cockroaches operate in essentially the same way. When a natural mechanism departs from what would seem to the investigator in his naïveté to be the most obvious and economical mode of operation, then there is usually some extremely interesting and novel reason waiting to be revealed. I shall speculate further on this and on other aspects of endogenous activity in a later chapter, but what is most needed at the present time is more definitive information on the behavioral function of the ocellus.

The Adaptive Value of Endogenous Activity. These few examples of endogenous activity in neurons found within the central nervous system and in receptor organs of insects serve to illustrate a very widespread phenomenon. In fact, the ubiquity of endogenous activity of nerve cells suggests that it may be the rule rather than the exception. This leads to the question of its adaptive value as compared with straightforward stimulus-response signaling.

Pumphrey[30] suggests that the extremely small amounts of energy needed to activate lateral line receptors and cochlear sense cells of vertebrates could be accounted for by assuming endogenous activity in the receptor cells. He points out that an endogenously active receptor has no fixed threshold, but swings from complete inexcitability during its discharge with concomitant refractory period to complete excitability just before the next discharge. At intermediate points in this cycle progressively smaller external stimuli could precipitate a discharge later and later, until the effective stimulus becomes zero at the moment of endogenous discharge. Reference to the analogy of the bucket (Fig. 39) will make this clear.

The intrinsic time constant of this cycle of excitability determines the frequency of the endogenous impulses. Interpolation of brief or continuous stimuli on this cycle would cause an increase in discharge frequency proportional to the stimulus magnitude. Stimuli that are vanishingly small must still have a corresponding effect on the frequency, but the point is that they are still proportionally effective rather than mere triggers. On the other hand, sitmuli that are subthreshold for a receptor that responds only when stimulated must be said to be completely ineffective since they fail to generate an impulse, the only transmitted signal that contains information as far as the central nervous system is concerned. Therefore, endogenous activity can be said to extend the effective range of a receptor cell to the level of thermal noise.

This extension of the range of sensitivity to its theoretical limit without narrowing the response range is obtained at the price of a tendency toward randomness in the sensory input. Presumably, this is counteracted in the central nervous system by a process of temporal summation of afferent impulses at the next neuron that determines behaviorally significant frequencies, as well as statistical equalization through spatial summation of the activity generated by a number of receptor units exposed to the same conditions.

Endogenous activity can increase the versatility of receptors in another way. If a certain rate of firing, say 5 impulses per second, is the endogenous frequency of a given receptor, this must signal zero stimulation from the outside. A stimulus, as ordinarily understood, would be expected to increase this frequency in proportion to its intensity. However, it is quite possible to visualize a reversal of the sign on the stimulus that would decrease the frequency from the endogenous rate. The situation is analogous to the action of EPSPs and IPSPs in causing the membrane potential of a motor neuron to deviate either toward or away from the threshold value (Fig. 37). An analogy for stimulus-response and endogenously active neurons is to be found in two types of electric meters, one having a zero point at one end of the scale, and the other with the zero in the center. The latter has the advantage of being able to measure current flowing in either direction, although some of its range is thereby sacrificed. In an endogenously active receptor the zero point is the endogenous rate of discharge, which can be either increased or decreased depending on the sign of the stimulus. This "null-point" type of receptor operation has been demonstrated in the statocysts or gravity receptors of lobsters. When gravity, operating through the otolith, produces no deviation of the hairs on the receptor cells, the latter discharge at a certain frequency. When the torque is in one direction this discharge increases; when it is in the other direction the discharge decreases.[8]

Keeping in mind the analogy of the electric meters, and remembering also that many receptors adapt during stimulation and disadapt in its absence, it will be realized that the latter have gained versatility in that they can change from one type of operation to the other depending upon the amount of stimulation to which they are exposed.

All of this endogenous activity is "as sounding brass, and a tinkling cymbal" as far as our present interest is concerned unless it can be shown to have some connection with what animals do. This is the topic of the following chapter.

10. *Endogenous Activity and Behavior*

The literature of ethology[23,57,58] and psychology contains much evidence that animals may behave like regenerative or relaxation oscillators. Familiar examples of endogenous behavior are connected with nourishment and reproduction. Relatively simple and nonspecific internal changes, such as alterations in the blood-sugar level or the concentration of sex hormones, are able to elicit restlessness or appetitive behavior. Under natural conditions this appetitive behavior, although apparently random in itself, increases the animal's chances of encountering stimuli enabling orientation toward food or mate. After eating or mating the appetitive behavior disappears, to return at some later time under the influence of appropriate internal changes.

If the animal fails to encounter optimal stimuli orienting it to food or mate the appetitive behavior continues, and the consummatory act of eating or mating is released by less and less "normal" substitute stimulus objects. There is considerable debate about what the animal is trying to do, and about the biological significance of the concepts of satisfaction, need, and goal. However, my present concern is merely to point out that in outline the cycle of events in appetitive behavior is analogous to the changes in excitability that take place in single neurons.[37] This suggests that something might be learned about neural mechanisms underlying some forms of appetitive behavior by searching for correlated endogenous activity in the nervous system.

Before examining this possibility in more detail it is worth while to speculate briefly on the general biological significance of endogenous or self-exciting systems. In Chapter 1 living and nonliving systems were considered in relation to the second law of thermodynamics. Living matter is able to store in chemical bonds some of the energy released by nonliving matter, and thereby to travel counter to "time's arrow." It was concluded that this was possible because the ability to store information in genes, nervous systems, and human records has enabled living matter to insulate itself in some degree from the ravages of time, and thereby to generate order in some parts of the universe at the expense of greater disorder in others.

Another attribute of living matter, or perhaps another aspect of this same attribute, is suggested by the fact that living things are invariably and continuously changing their position or form under conditions that are seemingly optimal and uniform.[37] We take it for granted that an animal or plant living in a uniform environment will continue to grow or move in the presence of a smooth energy flow, while a stone in the same environment will not. Stated more generally, living matter is always in a state of imbalance with its environment, while nonliving matter is always in a state of balance. This state of continual imbalance of living matter, of never reaching equilibrium in a physical sense, seems to be a general form of endogenous instability.

Animal behavior seems to be but a special example of the imbalance of living systems. Substrates of rhythmic activity modulated by stimuli from the outer world are common to protozoa, neurons, cockroaches, and men. Therefore, rather than regarding endogenous behavior as a paradox in need of special explanation, it seems to me to be more reasonable to regard it as another example of a general property of living matter. Until we know more about the nature of life it is about as appropriate to ask why animals show endogenous behavior as it is to ask why stones do not. If many find this to be an irritating and unsatisfactory conclusion, I hope they will make every effort to find another.

Whether or not we accept the presence of endogenous behavior as inevitable, it is of great interest to search for its mechanism and possible origin in the endogenous activity known to occur in the nervous system. The methods chosen will depend upon how the term is defined.

The term "endogenous" is applied here to behavior in the same sense that it was applied to neurons (page 144). Endogenous behavior is considered to have its origin in the intrinsic instability of some part of the central nervous system, and not as a response to specific stimuli reaching the organism from without. Stimuli acting through the sense organs may modulate endogenous behavior, but they do not trigger it.

Behavior of this kind should continue during complete sensory deprivation. Attempts to achieve this condition have been of two kinds. Complete elimination of all sensory input has been attempted by cutting all sensory nerves or by removing all sense organs, while leaving intact the connections between the central nervous system and the muscles. This method has not led to satisfactory conclusions because of the severe surgical insult produced by the operation. In the second method, the object is to isolate the animal from external stimuli. Observations of men and of animals while they are shielded from most external sources of stimulation for extended periods indicate that there is no over-all diminution in behavioral activity, but these experiments do not fully answer the question because the most ubiquitous of all earthly forces—gravity—cannot be eliminated for any length of time. A well-planned satellite experiment might eliminate all external sources of stimulation, although the subject would still be able to stimulate himself through his own actions.

Another view of the mechanism of endogenous behavior is suggested by the spinal-reflex mechanism discussed in Chapter 8. This is that the postulated state of internal instability is a product of internal factors such as blood chemistry plus the total sensory input acting in a nonspecific manner through the reticular formation or some analogous mechanism. In this case behavior should be absent or markedly altered in the sensory-deprivation experiments mentioned above.

Neither of these possibilities has been tested unequivocally because of the technical difficulties of achieving complete sensory deprivation. Most of the evidence, though indirect, suggests strongly that some forms of behavior can occur in the absence of sensory input. Von Holst in a series of papers[16] showed that the rhythmic movements associated with locomotion in fishes and earthworms are generated primarily by events occurring within the central nervous system. Wilson has shown[60] that the rhythm of nerve impulses that determines the basic wing-beat frequency in locusts arises endogenously within the thoracic ganglia. Another piece of indirect evidence is provided by the sexual behavior of the praying mantis.[32,46]

Mating in the Mantis. Mantids are relatively inactive insects. The cryptic coloration and form of most species gives them a close resemblance to the leaves, twigs, and bark of their natural habitats, where they wait in ambush for passing insects. Slight movement of an insect within reach of the specialized forelegs releases the rapid strike (Fig. 33). Insects up to the size of the mantis may be caught and eaten, including other mantids of the same species.

This fortuitous cannibalism is probably minimized in nature by the inactive and cryptic habits of mantids, and by their tendency to scatter. It is interesting to note that during the first nymphal stage, just after dozens or hundreds of individuals have hatched from a common egg case and are necessarily crowded, there is considerable locomotor activity and no sign of cannibalism. The inactive habit and the tendency to strike at any moving object of an appropriate size become established only after the young mantids have scattered. However, the cannibalistic habit seems to become a menace to racial survival during the sexual encounter between male and female.

Courtship is similar in the European Mantis (*Mantis religiosa*)[32] the Chinese Mantis (*Tenodera sinensis*), and the Ceylon Mantis (*Hierodula tenuidentata*). To our eyes it resembles a sneak attack rather than a courtship. The mature male is slenderer and much smaller than the female (Fig. 42). He has larger compound eyes and ocelli, and longer antennae, and is somewhat more active. The male freezes on seeing the female, a reaction sometimes shown by both sexes to large moving objects. He may remain motionless for hours, or he may begin

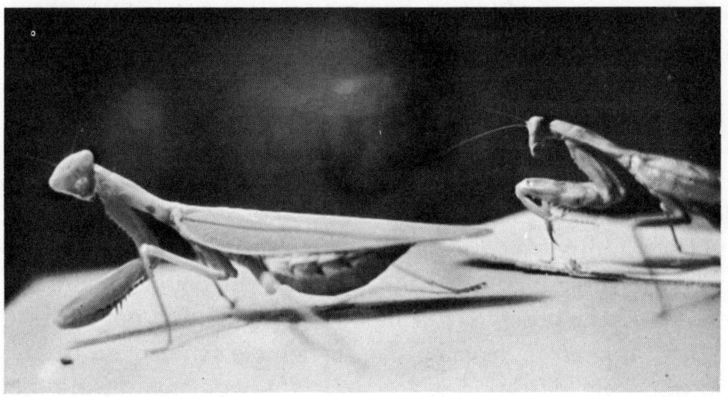

Fig. 42. Male and female praying mantis (*Hierodula sp.*). The male is about to mount the female. Note the longer antennae and slenderer and smaller build of the male.

a slow and measured approach, stalking her always with his eyes fixed upon her. Movements by the female cause him to freeze once more, so that this approach may take minutes or hours, during which time the male shows no other signs of sexual excitement. When the pair are within a length of each other the male mounts the female by taking a short flying leap, landing usually head-to-head, but sometimes head-to-tail. In the latter event he turns rapidly after a few moments, and clasps the female with his forelegs (Fig. 43*A*).

Once in this position, the male intermittently flagellates the female's head with his antennae, and begins to make copulatory movements. The abdomen assumes an S-shaped position that twists the terminal segments through a right angle and directs them forward, from which position the male genitalia repeatedly probe the ovipositor of the female. Coupling is accomplished after 5 to 30 minutes by separation of the ovipositor valves, and continues for several hours during the formation of a spermatophore in the female genitalia.

The female often makes no obvious responses to the male throughout these activities. She may flatten her body slightly upon being mounted, a movement that also seems to be part of the cryptic posture assumed on the approach of a large predator. It is probable that her active participation is required during coupling; otherwise she appears to be passive as far as the sexual aspects of the encounter are concerned. However, she may strike at the male and begin to eat him.

This cannibalistic attack may take place during the male's approach, directly after mounting, or as the couple separates. It has not been observed during actual copula. It is not inevitable, and appears to depend upon the female's state of nutrition, and upon her detection of the male's movements during the final phase of his approach. It is less likely to occur if the female is feeding, and if the male approaches from the rear. A few males have been observed to mate a number of times with

Fig. 43. Sexual behavior and cannibalism in *Hierodula*.[46] The insects normally hang upside down. (*A*) Copulating pair in which the male has not been attacked. (*B*) Female consuming the head of the male; the male abdomen already shows the bending movements of copulation. (*C*) Decapitated male copulating.

the same or with different females without losing their heads, while many succumb on the first encounter.

The head and prothorax of an approaching male are naturally most exposed to the female's attack, and are therefore eaten first (Fig. 43*B*). This is followed within a few minutes by intense and continuous sexual movements (S-bending) of the male abdomen accompanied by curious locomotor movements that have never been observed in the intact male mantis. These are lateral or rotary walking movements that tend to carry the body of the male from the head-to-head position at the time of the attack to a position parallel to the female, and eventually onto her back. The movements are sufficiently powerful that the remains of the male's body are pulled out of reach of the female mandibles, and she is rarely able to eat more than part of the male thorax. The sexual movements of the male abdomen continue with vigor, and coupling takes place shortly after mounting (Fig. 43*C*). Copulation continues for several hours and a normal spermatophore is formed.

Surgical decapitation or transection of the nerve cord at any point releases intense sexual behavior in males, even before sexual maturity.[32] Rotary locomotion and S-bending of the abdomen continue for several days, and the decapitated male clasps any object the approximate size of the female body and makes vigorous and prolonged attempts to copulate. Although a decapitated male cannot locate a female, he clasps her immediately if placed on her back, and normal copulation usually follows. This reaction was put to practical use in maintaining an inbred strain of *Hierodula* in which the intensity of courting in intact males was much reduced. A high percentage of fertilized eggs was realized by decapitating the males and placing them on the females.

Enhanced sexual behavior following decapitation has been observed in other insect species, notable in females where egg laying may be released by decapitation or by separation of the abdomen from the thorax.[24] Headless female mantids make continuous movements of the ovipositor, and often deposit ill-formed egg cases. It seems that this behavior, as well as the sexual behavior shown by the decapitated male mantis, must have adaptive value in terms of species survival, tending to offset attacks by predators and the cannibalism of the female mantis. This makes it worth while to inquire whether a special neural mechanism has evolved in this connection.

It is clear that in addition to nerve centers in the head many important sense organs, notably the compound eyes, ocelli, and antennae, have been eliminated by decapitation. However, removal of these sense organs severally or collectively causes no increase in sexual behavior. Also, most of the brain can be removed with no change in the sexual activity, and it is only after elimination of the subesophageal ganglion—a mass of nerve tissue whose other functions include innervation of the mouth parts (see Fig. 56)—that continuous sexual behavior and rotary locomotion is released. It appears that nerve impulses are continuously transmitted from this region down the nerve cord to local centers, where they inhibit the regional motor activities associated with mating. In the intact male this inhibitory effect is counteracted in some manner only upon actual contact with the female.

This implies that the local patterns of nerve activity responsible for rotation and sexual movements are endogenous and autonomous, being determined by built-in neuron interactions within the ganglia that serve these segments of the body. As mentioned above, it is practically impossible to test this hypothesis directly by removing all sensory input in order to observe whether behavior continues in its absence. Most insect nerves contain both sensory fibers running from the periphery and motor fibers that transmit impulses to the muscles, so that transection of all nerves would not only eliminate the sensory input, but also make it impossible for the behavior to take place. However, electrophysiological methods[46] have provided a fairly conclusive indirect test of the endogenous nature of the sexual movements made by the male abdomen.

A male mantis was restrained on a cork platform, and its abdomen was opened from above so as to expose the nerve cord. The terminal ganglion of the ventral chain (Fig. 56) supplies the sense organs and muscles of the posterior abdominal segments concerned in the S-bending and copulatory movements of the phallic apparatus. The nerves connected with these activities fan out from the ganglion in all directions, and are readily accessible (Fig. 44). All of these nerves were cut, and electrodes were placed on the central stump of one known to supply the muscles of the phallic apparatus. Following this operation there was no possibility of impulses from local sense organs reaching the ganglion, except via the connectives joining it to the rest of the nerve cord. At the same time, the electrodes provided a means for measuring the output of motor-nerve impulses that would have caused contractions in the phallic muscles under normal circumstances. A sample of this motor-impluse pattern is shown in Fig. 45.

This motor activity was observed on the oscilloscope for a sufficient time, perhaps 30 minutes, to determine its average level. Then the nerve cord was severed just above the terminal ganglion, eliminating the last possible avenue whereby sensory impulses might have reached the ganglion, and at the same time preventing the arrival of the inhibiting impulses from the subesophageal ganglion. This injury caused a brief flurry of impulses, but the motor activity continued at its normal level for a few minutes. After 1 to 5 minutes the spike activity began to mount and continued to do so for 10 to 15 minutes until the number of motor impulses had increased manyfold (Fig. 45). The increase was due in part to the onset of activity in a number of fibers (large spikes) that were previously inactive when the nerve cord was intact, and in part to increases

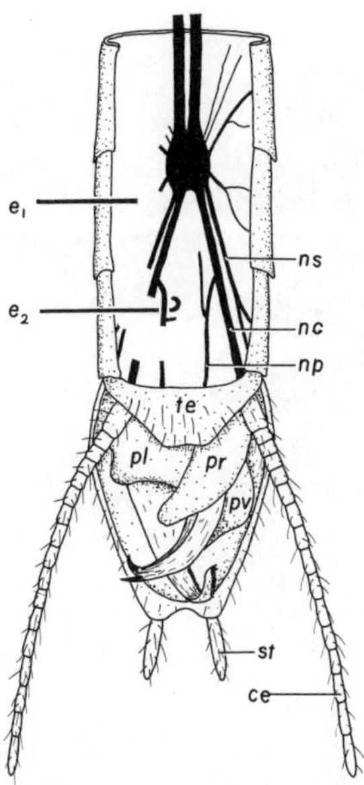

Fig. 44. Terminal segments of the abdomen in the male *Mantis religiosa*. Viscera and testes have been removed. Nerves passing to the last abdominal ganglion have been severed on the left side, and an electrode e_1 has been placed under the proximal stump of the phallic nerve. In an actual experiment on endogenous activity the nerves on the right side would also be severed. The indifferent electrode e_2 is in the body fluids. *ce,* cercus; *nc,* cercal nerve; *np,* phallic nerve; *ns,* stylus nerve; *pl, pr, pv,* complex and asymmetric phallomeres employed in coupling with the female; *st,* stylus. (Modified from Roeder, Tozian, and Weiant.[46])

in the frequency of impulses in fibers that had previously fired only occasionally.

This increased activity of motor neurons in the terminal ganglion continued for several hours, or for the life of the nerve preparation. Often there were interesting and complex patterns of waxing and waning, individual fibers increasing and decreasing their discharge frequency several times a minute. Different units had different rhythms, and simultaneous activity recorded from other motor nerves from the ganglion

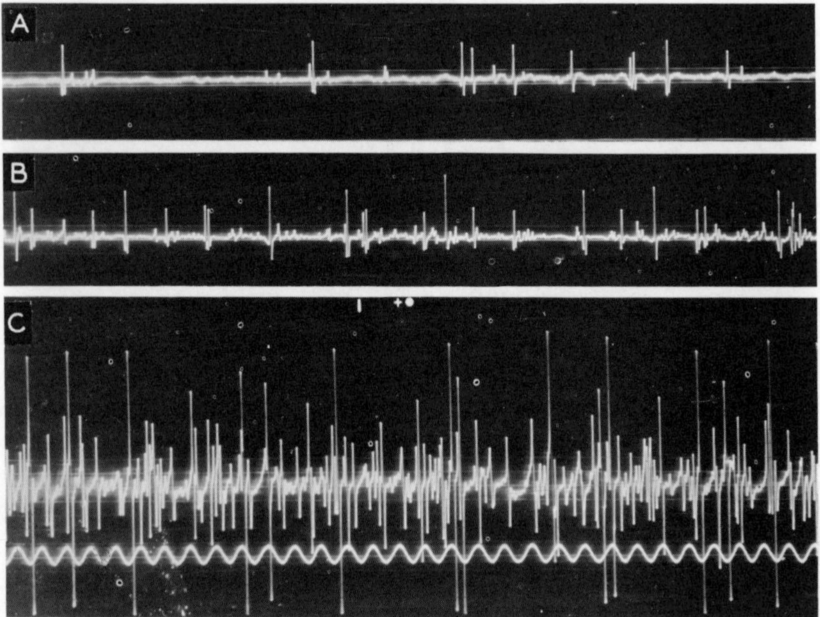

Fig. 45. Pattern of motor-nerve impluses in the phallic nerve of *Mantis religiosa.*[46] The phallic nerve has been severed distal to the electrode and all other nerves connected to the ganglion have been cut: (*A*) before connectives between the last abdominal ganglion and the rest of the nervous system have been severed; (*B*) 3 minutes after transection of the nerve cord; (*C*) 7 minutes after transection of the nerve cord. In (*B*) and (*C*) the ganglion has been completely isolated from sensory input. Compare with Fig. 37. Time line in (*C*), 100 cy/sec.

showed still other unsynchronized rhythms. Since the nerves concerned supply various muscles of the phallic apparatus and abdominal wall, it is hard to avoid the conclusion that these organs would have been making rhythmic movements had they still possessed their normal connections. Furthermore, the conditions of the experiment leave no doubt that this rhythmic activity in motor nerves was endogenous.

This experiment seems to show as directly as possible that a behavior pattern of adaptive value in survival of the species has its origin in an autonomous built-in neural mechanism. However, it does not prove that the patterns of motor impulses released by transection of the nerve cord would have produced coordinated and effective movements of the sort appearing in the decapitated but otherwise intact male mantis, so that it is not possible to estimate the steering function of sensory impulses from the genitalia and surrounding regions when superimposed on this endogenous background.

Endogenous Activity in the Cockroach. One might assume from this that an endogenous neural mechanism has evolved especially in response to the cannibalistic habits of the mantis. However, the observations of endogenous activity in isolated nervous systems (Chapter 9), and behavioral experiments with various insects, suggest that there is a general tendency of local neural mechanisms to be endogenously active, and that the situation found in the male mantis is merely a special case in which this endogenous tendency has become modified to serve a function of adaptive value to the species. This conclusion is strikingly illustrated by similar experiments carried out with cockroaches.

These experiments were begun originally for purely expedient reasons. It is both expensive and difficult to maintain a year-round colony of mantids and a continuous supply of living insects upon which to feed them. Because of their cannibalism, mantids must be raised in individual containers, and the smaller males are invariably in short supply. At one period when the colony was at a low point we were forced to make shift with the ever-present and reliable cockroaches in carrying out the neurophysiological experiments. These experiments became so interesting in themselves that they merit separate discussion.

The terminal ganglia of cockroaches of both sexes were exposed, and all the nerves cut. Electrodes were placed on the central stump of a branch that originally supplied the genitalia, and motor-impulse activity was measured before and after decapitation or cutting the nerve cord. The results were essentially similar to those obtained with the male mantis. Within a few minutes after complete isolation of the terminal ganglion the endogenous activity of motor neurons began to climb steadily, and it continued at a high level for several hours. The activity of individual units waxed and waned in a complex pattern several times a minute (Fig. 46), each unit following a pattern of its own. As in the mantis, this activity could be released either by cutting the nerve cord or by decapitation.

This led to similar experiments in isolating other ganglia,

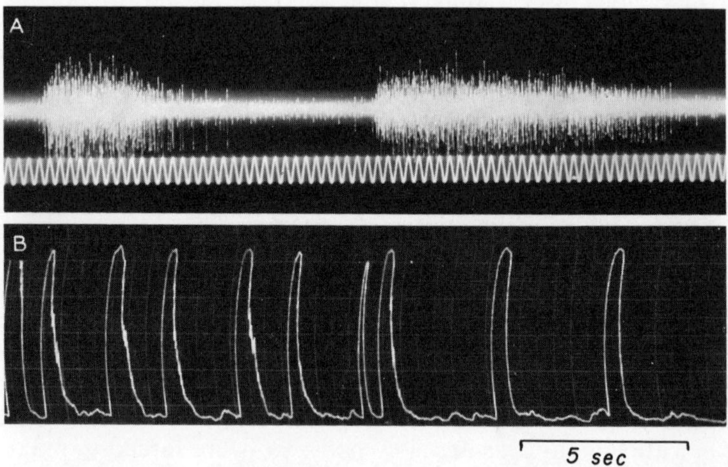

Fig. 46. Endogenous activity in nerve *X* following transection of the abdominal-nerve cord in the cockroach *Leucophaea maderae.* (*A*) Two bursts of efferent impulses; note that the time scale (10 cy/sec) is contracted compared with that in Fig. 45. (*B*) Efferent activity electronically integrated and recorded simultaneously from the same preparation; the time scale has been still further contracted. The short and long bursts near the middle of the integrated record correspond to those shown in (*A*). (Milburn, unpublished).

particularly the large thoracic ganglia supplying the legs.[66] Here also it was found that the activity of many motor neurons was partially or completely suppressed by connections with other parts of the nervous system, and even with the opposite side of the same ganglion. The greater the isolation from other nerve connections, the greater the tendency to increased frequency of firing. This suggests that a similar motor mechanism, normally suppressed by impulses from higher centers, may be responsible for the curious rotary behavior in headless male mantids, although this has not been directly studied.

The discovery of endogenous motor activity in the cockroach led us to examine the sexual behavior of this insect for comparison with that of the mantis. Mating behavior in a number of cockroach species has been studied extensively by Roth and Willis.[51] It is very different from that of the mantis, search behavior in the male being triggered by an odor released by virgin females. Details of the approach and coupling need not concern us here, except to note that cannibalism does not occur and the male is at no time in jeopardy. Therefore, the neural mechanism that appears to have value in racial survival in the mantis can have no such significance in the cockroach.

Adult male cockroaches were then decapitated and their subsequent behavior was observed. Such insects survive for several days but show little or no locomotor activity. They often stand with the body raised higher than normal above the ground, and occasionally clean the abdominal tip with their hind legs. Closer observation of the abdomen showed that various small movements take place. These include repeated raising and lowering of the abdomen and slight peristaltic movements of the terminal abdominal segments. The last segment, bearing the genitalia, is sometimes turned upward and to one side, this action being accompanied by movements of the phallic organs and the cerci. Some of these movements have been observed in sexually excited intact cockroaches, but after decapitation they appear to be merely uncoordinated fragments of the sexual pattern, and are completely ineffective in securing copulation. The bodies of decapitated males were repeatedly brought into contact with receptive virgin females, but copulation could never be induced.

While there is little doubt that these movements are a manifestation of endogenous activity released in motor neurons of the terminal ganglion by separation from the inhibitory center in the subesophageal ganglion, it seems equally certain that they are of no adaptive value to the species. The only explanation seems to be that the general plan of the insect nervous system is based on a series of locally autonomous and endogenously active centers that are coordinated to subserve the whole through higher levels of inhibitory control. In the male mantis, and perhaps in female insects that oviposit after destruction of the head and thorax, this basic arrangement has come secondarily to have adaptive value in species survival. We must seek its origins elsewhere (Chapter 13).

Before turning to general questions about the organization of the insect nervous system it is worth while adding a postscript to the experiments on the terminal ganglion. Irrespective of whether endogenous activity of neurons in the terminal ganglion is the basis of adaptively significant behavior or not, there must be other and less irrevocable ways of releasing it. Some male mantids are able to mate several times without losing their heads, and the Humpty-Dumpty expedient of decapitation used in the experiments is technically unsatisfactory because of its irreversibility. Various attempts to cause a temporary block in the inhibitory pathways from the head met with no success until a shot-in-the-dark experiment[26] was suggested by some observations[28] that extracts of the corpora cardiaca injected into cockroaches caused changes in their locomotor behavior.

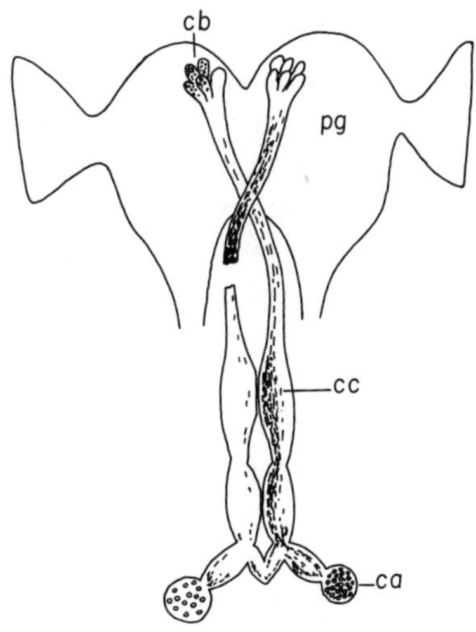

Fig. 47. Dorsal view of the corpora cardiaca of the cockroach *Leucophaea maderae.*[55] The cell bodies *cb* of the corpora cardiaca *cc* lie in the pars intercerebralis (see Fig. 59) of the protocerebral ganglia *pg*. Their axons become the enlarged body of the gland, which terminates in the corpora allata *ca*. Surgical transection of the connection between cell bodies and gland on the left has caused accumulation of secretion in the proximal stump of the nerve, absence of secretion in the body of the gland, and hypertrophy of the corpus allatum.

The corpora cardiaca are large glands lying just behind the insect brain and in the dorsal region of the neck (Fig. 47). In cockroaches they are bluish translucent sacs apparently formed by the grossly inflated axons belonging to certain neurons lodged in the brain. These axons appear to have departed from their conventional role of acting on other neurons or muscle fibers by releasing a mediator substance over submicroscopic distances, and instead to have been converted into neurosecretory reservoirs.

The corpora cardiaca were dissected from two or three cockroaches, ground up in a small mortar with salt solution, and applied directly to the subesophageal ganglion or nerve cord of an insect prepared for recording the motor activity of the thoracic or terminal ganglia.

Five or 10 minutes after this treatment axons in the phallic nerves began to discharge with increasing frequency (Fig. 48) and after about 30 minutes motor activity was similar in level and pattern to the endogenous rhythms that followed decapitation or isolation of the ganglion. Rhythmic bursts of spikes continued for an hour or so, but eventually declined, leaving motor activity at a level typical of that in an intact or untreated ganglion. A second application of extract caused this cycle to repeat, and rhythmic bursts could be maintained for several hours by repeated applications. The active principle in the extract was rendered more stable by warming to boiling point, and frozen extracts could be stored for several weeks. Similar effects have not been obtained with extracts made from other organs of the cockroach, or with a number of natural or synthetic chemicals known to be active on neurons. The extract produced similar effects on motor neurons in the thoracic ganglia, but its action on internuncial neurons or the cercal-

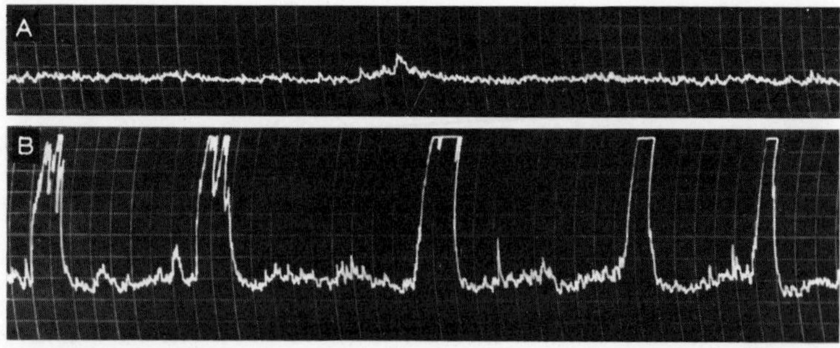

Fig. 48. Integrated record of motor nerve activity from the last abdominal ganglion of the cockroach *Periplaneta americana*. Connections with the subesophageal ganglion were intact throughout. (*A*) before; (*B*) 14 minutes following the application of an extract of the corpora cardiaca of *Leucophaea maderae*. (Milburn, unpublished).

nerve giant-fiber synapses (Chapter 7) was either equivocal or negative.

Owing to the expenditure of time and cockroaches needed to collect and extract the number of corpora cardiaca (each smaller than a pinhead) needed for more detailed chemical studies, it has not yet been possible to isolate the active principle. What little we know hints that in inhibiting some of the systems that in turn inhibit the local endogenous rhythms of segmental ganglia this principle plays a role intermediate between that of the diffuse and general spread of hormones and the specific local action of synaptic mediators.

We have encountered a combination of endogenous neural activity and inhibition at two widely separated and functionally distinct points in the neural mechanism of behavior—the ocellar and the reproductive systems. The significance of this in the functional organization and origin of the insect nervous system will be discussed in Chapter 13.

11. Inside a Moth

In an earlier chapter it was pointed out that the simple nature of the nerve signals transmitted inward from the pair of *A* cells in a moth's ear invites an effort to investigate the neural machinery used by moths when they respond to the cries of bats. But at that point the "black box" of the moth's central nervous system was left unopened, and attention was focused on defining its input–output relations (Chapters 5 and 6). Subsequent chapters examined the central nervous systems of other animals in an attempt to learn more about the properties of the components hidden within and to discover ways of measuring their performance. Now it is time to inquire what happens to nerve signals from the *A* cells after they enter a moth's central nervous system.

Turning-Away. First it is necessary to look more closely at the simplest and at the same time most interesting form of behavior shown by moths during exposure to ultrasonic pulses. Turning and steering a course away from a source of ultrasound was observed to occur in the field when a moth encountered ultrasonic pulses of low intensity (Chapter 5). Only one of the *A* cells could have been stimulated at this sound level, and it seems likely that the central nervous system contains a mechanism that discriminates or compares *A* signals such as those shown in Fig. 15, using criteria of the *A* response from among those listed on page 47. The decision reached by this mechanism on receiving each cry uttered by a still-distant bat must repeatedly operate the steering mechanism through negative feedback, so as to keep the moth on a course away from the predator. How much time is occupied by this central discrimination, and how does a moth accomplish the turning-away movement?

Precise answers to these questions cannot be obtained by watching a moth flying free. The experimenter must come to closer grips with his subject in the laboratory, while realizing how much he is interfering with its natural behavior.[45]

Wild moths were captured in an ultraviolet light trap. Each was attached to a fine insect pin by a small drop of wax applied to the thorax in such a way that this restraint did not interfere with the flapping of its wings. A certain number could then be induced to fly continuously when their feet were lifted out of contact with the ground. Their flight performance was measured in two ways. The pin bearing the moth was inserted into a phonograph pickup in order to register the vibrations of the thorax as electrical signals, and the flying moth was placed in front of a differential anemometer that indicated the direction of the aerial wake thrown back by its beating wings (Fig. 49). The anemometer consisted of two minute thermistors, one placed behind the wings on the right side and the other behind those on the left. Thermistors show large changes in electrical resistance when they change in temperature. These thermistors were heated electrically and the resistance of one relative to the other was continuously recorded. When the same wind velocity impinged on both, there was no difference in their electrical resistances; when the moth attempted to turn to the right there was a relative decrease in the velocity of air propelled backwards by the right-hand pair of wings, a relative increase in the temperature of

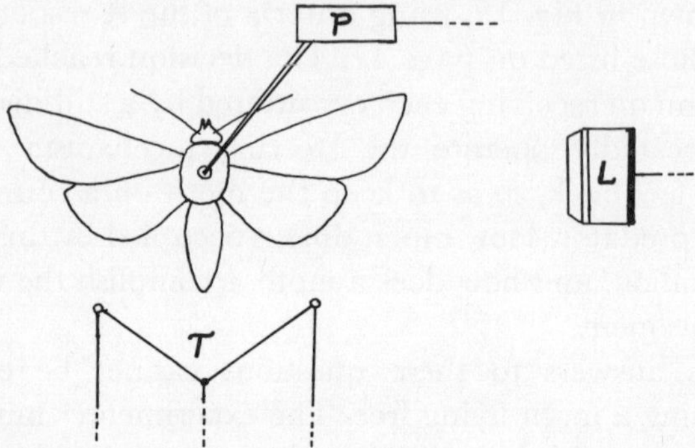

Fig. 49. Diagram of the arrangement used to register turning tendencies. The moth is attached by wax to a pin, which is inserted in a phonograph pick-up (*P*). The loudspeaker (*L*) is directed at it from one side. The pair of thermistors (*T*) is placed in the wake of the flapping wings. Signals from all three devices are led to an oscilloscope.[45]

the thermistor on that side, and a relative decrease in its resistance. The thermistors were included in a bridge circuit whose balance was registered on an oscilloscope trace. This signal indicated the latency and the direction of the moth's tendency to turn upon receiving an ultrasonic signal from one side or the other.

A loud-speaker emitting ultrasonic pulses on demand was directed at the flying moth from various angles. Signals indicating its operation were displayed on one trace of the oscilloscope record. The phonograph pickup record of thoracic vibrations and the anemometer record indicating turning appeared on two other traces.

Before examining the traces it is necessary to assess some of the limitations placed on the moth by being mounted in this apparatus. As is usual in any laboratory experiment, there are many restrictions that could not be evaluated but undoubtedly accounted for the fact that only about half the moths tested could be induced to fly under these conditions, yet those that refused appeared to be uninjured and were found to fly perfectly well when demounted and released out-of-doors after the experiment. A still smaller percentage of the "flyers" showed responses to sounds from the loud-speaker, and the only comfort is that some of these could be induced to perform consistently over and over again.

An important experimental restriction is due to the fact that the moth is fixed in position relative to the angle of the sound source, and its efforts do not change its flight path. Therefore one cannot speak of the moth turning away but only of its *turning tendency*. If it were free to turn the change in pitch of its wing strokes would presumably swivel it relative to the sound source, and further steering corrections would keep it on a bearing away from the latter. In our apparatus the moth's efforts have no effect on its bearing and merely bring about a change in the wake it generates. From the moth's viewpoint it must seem that the "bat" is flying round it or that there is no response to its "rudder," and this might be expected to cause it to make still greater efforts to turn. Therefore, the anemometer records can give reliable information on the time when the turn was begun and on its sign—to the right or left—but not on the magnitude of the attempted turn.

Records of these turning tendencies are shown in Fig. 50. They fully confirm the earlier observations of free-flying moths (Chapter 5) and make possible a much more precise evaluation of what happens. The attempt to turn away from the sound source is seen to begin 50–55 milliseconds after the first sound pulse. When certain corrections are applied to the measurement the minimum reaction time turns out to be 40–45 milliseconds. This is much shorter than times measured from the photographic tracks of free-flying moths, and very similar to other animals' minimum startle-response times (Chapter 7). Since the sound pulses serving as stimuli are 100 milliseconds

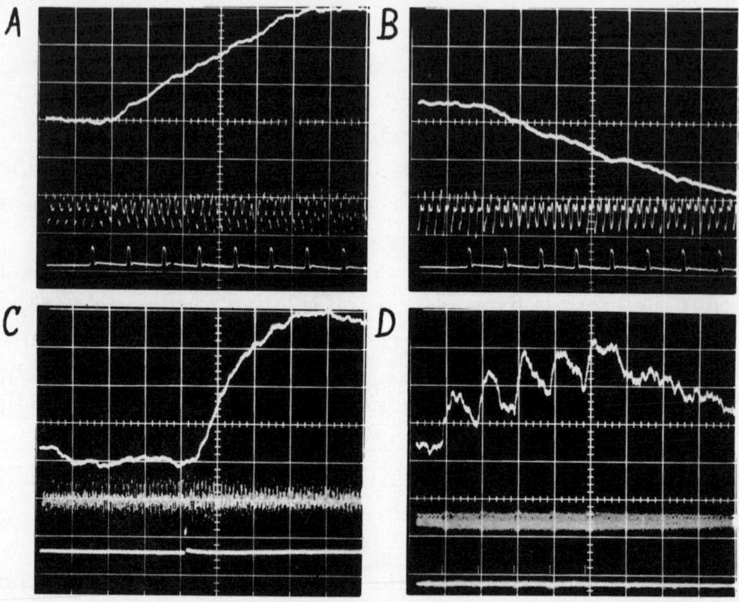

Fig. 50. Records of attempts made by moths to turn away from a source of ultrasonic pulses placed in the horizontal plane and at right angles to the body axis. An attempted right turn is indicated by an upward deflection of the upper trace. Middle trace, thoracic vibrations; lower trace, signal indicating stimulus pulses. (*A*) *Leucania* sp., stimulus from left side. (*B*) The same, stimulus from the right. Stimulus was 40-kc/s pulses 10 msec in duration recurring at 10/sec and roughly +20 db above the threshold of the tympanic organ. Vertical lines mark 0.1-sec intervals. (*C*) *Agrotis ypsilon*, response to a single 10 msec 40 kc/s pulse at +40 db originating at left side of moth. Vertical lines mark 0.2 sec intervals. (*D*) *Feltia subgothica*, single but summating turning responses to the same signal delivered at 2-sec intervals. Vertical lines mark 2.0-sec intervals.[45]

apart, this means that the decision made to respond as well as the choice of the correct direction are arrived at in the moth's central nervous system on the basis of a comparison of the spike trains generated in the right and left *A* cells by a single sound pulse. This is shown clearly in Fig. 50*C*.

In other experiments using the anemometer it was found that if the right tympanic organ was destroyed the moth would turn to the right—to the "silent" side—irrespective of the side of its body encountering the most intense sound pulse. Indeed, it developed a flight bias in this direction even when no stimulus was presented, suggesting that even the endogenous activity of the *A* cells in the intact ear is able to operate the steering mechanism when this activity is not counterbalanced by endogenous activity from receptors on the opposite side.

The simplicity and sharpness of the turning records suggested an experiment to see what a moth does when it attempts to turn. A camera was mounted above the moth and anemometer and the area illuminated by a strobe lamp. The shutter was opened for ⅛ second at a time when the anemometer showed the moth to be making a maximum effort to turn. The scene was illuminated by seven or eight short flashes from the strobe lamp while the shutter was open. This gave a picture showing multiple images of several wing positions assumed by the moth during its turning attempt (Fig. 51). It can be seen that the moth behaves just like a terrestrial or aquatic animal making a similar movement. Its antennae, head, and body are all bent into the turn. The turn is attempted by partially folding the wings on the inside while those on the outside of the turn appear to be extended more than before. The moth may be said to feather one propeller while increasing the pitch of the other, thus attempting to yaw itself into the turn. There appear to be no consistent changes in thoracic vibration while the moth is trying to turn, so that the main muscle motor that flaps the wings does not seem to be concerned. The motor agents in turning are probably the direct flight muscles—certain small muscles attached to the wing base that are responsible for changes in wing angle during flight.

These studies with the anemometer greatly narrow our search for internal mechanisms. They tell us that the total

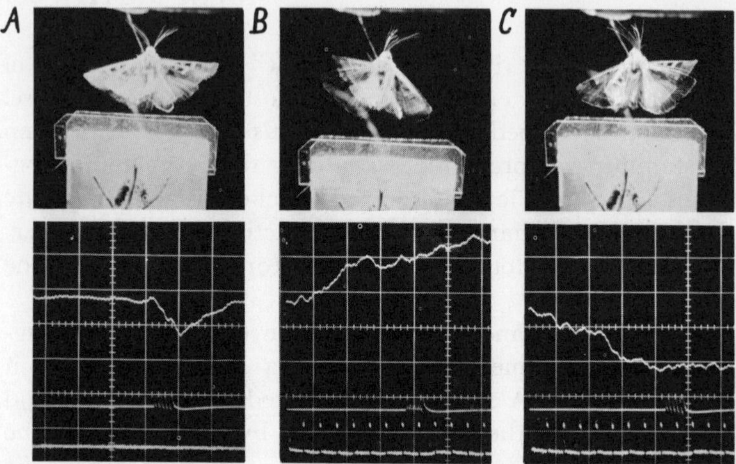

Fig. 51. Actions made by *Feltia subgothica* in attempting to turn away from a sound source. Multiflash pictures of moth are shown above a simultaneous anemometer record of its wake (upper trace). Time during which the camera shutter was open is shown by deflection in middle trace. Lower trace shows series of ultrasonic stimuli. The first sound pulse coincides with the beginning of the traces. (*A*) No stimulus. The sounds made by the shutter and flash caused some perturbation of moth's flight pattern. (*B*) 40-kc/s pulses 10 msec in duration at 10/sec and +30 db re the tympanic threshold coming from the moth's left side. (*C*) The same stimulus coming from the right side.[45]

neural circuit operates within 45 milliseconds from the arrival of a sound pulse at the tympanic membrane to changes in the direct flight muscles. Furthermore, this requires the arrival of only a single faint ultrasonic pulse, that is, a single spike train, from the more sensitive *A* receptor cell in the right and left ears.

Probing for Acoustic Interneurons. The anemometer records in Fig. 50 present turning-away behavior reduced to a very simple form. Yet, a series of complex transformations must have occurred in the moth's central nervous system after the arrival of a sound pulse at its ear and before its wake changed direction. Figure 50*C* shows how a single ultrasonic pulse

0.01 second long can cause a change in wake direction lasting 2 seconds or more. What events take place during the 45 milliseconds between stimulus and response? How is an ultrasonic pulse transformed into a change in wing movement? How is the specificity and sign of the response determined?

These questions can be answered only by examining the synaptic interactions between internuncial neurons lying in the mesothoracic and metathoracic ganglia of the moth. These ganglia are the main sensory and motor centers for the second and third thoracic segments that bear the two pairs of wings. They are fused together in Lepidoptera and lie in the ventral region of the thorax (Fig. 52). They can be compared with

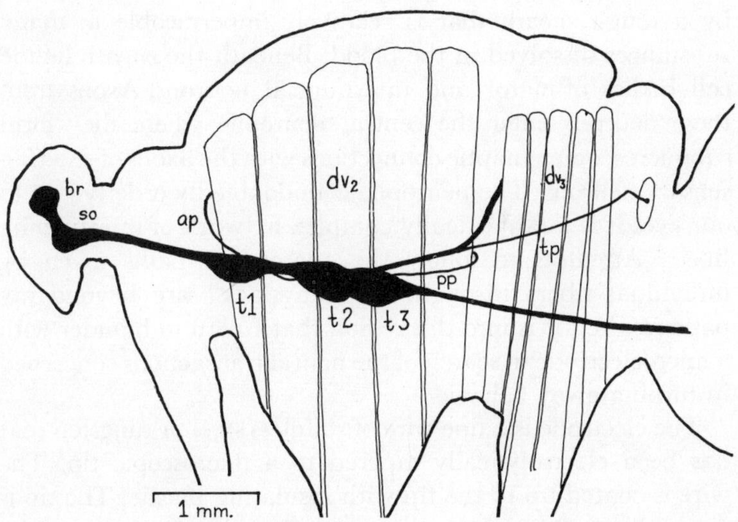

Fig. 52. Diagram of a moth's central nervous system (most of the muscles and nerves have been omitted): *ap*, anterior pleural nerve supplying some of the direct flight muscles that regulate wing angle; *br*, brain; dv_2, dv_3, dorsoventral flight muscles that raise the wings; *pp*, posterior pleural nerve; *so*, subesophageal ganglion; t_1, t_2, t_3, pro-, meso-, and metathoracic ganglia; *tp*, tympanic nerve coming from the tympanic organ and joining the main sensory nerve of the hind wing.

homologous ganglia in the cockroach (Fig. 31) and mantis (Fig. 56). The tympanic nerve enters the anterior dorsal region of the metathoracic ganglion in company with other sensory fibers from the hind wing. Motor nerves bound for the indirect muscles of the forewing (the main source of power during flight) leave the mesothoracic ganglion. In an average-sized moth the pterothoracic ganglion, as the fused meso- and metathoracic ganglia are often called, measures about 1 millimeter long and about 0.3 millimeter wide. The mechanisms we seek are packed into the mesothoracic portion, which has a volume of about 0.03 mm^3.

The fine structure of moth ganglia is similar to that of a cockroach ganglion (Fig. 38). The nerve tissue is surrounded by a tough sheath that is relatively impermeable to many substances dissolved in the blood. Beneath the sheath lie the cell bodies of motor and internuncial neurons. Axons from these neurons enter the central neuropile where they form submicroscopic synaptic connections with the axons of entering sensory neurons. The neuropile is undoubtedly orderly; yet to our eyes it is a fantastically complex feltwork of intertwining fibers. Attempts to follow the anatomical paths taken by individual fibers through this "haystack" are beyond my patience. Yet, it is into this region that I plan to blunder with a microelectrode in search of the neural transactions concerned in turning-away behavior.

The electrode is a fine wire of stainless steel or tungsten that has been electrolytically tapered to a microscopic tip. The wire is coated up to the tip with insulating plastic. The tip is gradually lowered into the neuropile at selected spots by means of a micromanipulator. The moth is restrained and prepared for the experiment in much the same manner used for recording activity in the tympanic nerve (Chapter 4). For ganglion-probing, however, the insect is mounted ventral side up, and the ventral cuticle and muscle is dissected away so as to expose the lower surface of the pterothoracic ganglion.

The nerve connections of both ears with the central nervous system are left intact. Small loud-speakers placed on either side of the moth make it possible to stimulate either one or both ears with electronically generated ultrasonic pulses of various intensities and recurring in various patterns.

Before going further it is necessary to pause in order to contemplate the electrical bedlam encountered by the microelectrode as it pushes its way through the ganglion. The electrode tip, though small, is not small enough to penetrate the interior of the minute nerve fibers that compose the feltwork of the neuropile. Electrically speaking, these fibers lie in a conducting medium, so that the electrode tends to pick up large spike potentials from nerve impulses invading fibers close to its tip and smaller spikes from impulses traveling in more distant fibers. If the amplified spike potentials are converted into clicks by means of a loud-speaker, or into deflections on an ocilloscope screen, a given source of spikes seems to wax and wane as the electrode tip is slowly advanced past it through the neuropile. At some spots there is a veritable blizzard of spikes—activity in one region sounds like frying bacon! But in regions where the traffic is less heavy individual spike rhythms can be made out and brought to their maximum amplitude by making small adjustments of the micromanipulator.

It is frustrating not to be able to read the messages conveyed by most of the neuron "voices" heard as the microelectrode is advanced through the neuropile. It is like walking through a packed and noisy crowd waiting at an international air terminal. As one passes by, each voice in turn grows and then is drowned out by the general crowd noise. One may guess that one speaker is excited and another calm, but the message and even the language is generally incomprehensible. One must withdraw to the edge of the crowd in search of a familiar voice.

In the moth's ganglion the familiar signal is the response

of the *A* receptor neurons to a regular ultrasonic pulse coming from one of the loud-speakers. This signal is rendered more obvious on the oscilloscope screen by allowing the stimulating pulse to trigger the sweep of the electron beam so that the resulting train of *A* spikes appears at about the same spot during the beam's progress across the screen. This sort of pattern is displayed in Fig. 10. We must first search for a similar stimulus-locked pattern of spikes after the *A* axons have entered the neuropile.

When such a pattern is detected the microelectrode is maneuvered near the unseen source until the spike pattern stands out against the general neuron noise (Fig. 53*A*). If the response has properties similar to those listed on page 47 the odds are that the tip of the electrode is close to central portions of the *A* axons. This is reassuring although uninteresting. We wish to listen to a novel neuron, yet one whose behavior is in some way related to the familiar *A* response.

Experiments such as those described below required many thousands of electrode passes through the neuropile of many dozens of moth ganglia. Electrode positions were related to surface landmarks and the depth of penetrations estimated, but in the main electrode placement was essentially blind and dependent on chance encounters—each adding perhaps another piece to the jigsaw puzzle. The puzzle is by no means complete, and I can only describe parts of it that seem related to one another or to the key piece—the incoming train of *A* spikes.[44]

The most frequently encountered stimulus-related signals are produced by interneurons that have been named *repeater* neurons. The repeaters relay almost the same spike pattern passed on to them by the *A* fibers. The difference between *A* and repeater activity is clear only when the electrode tip simultaneously encounters sources of both *A* and repeater spikes (Fig. 53*B*). The repeaters fire later and adapt more

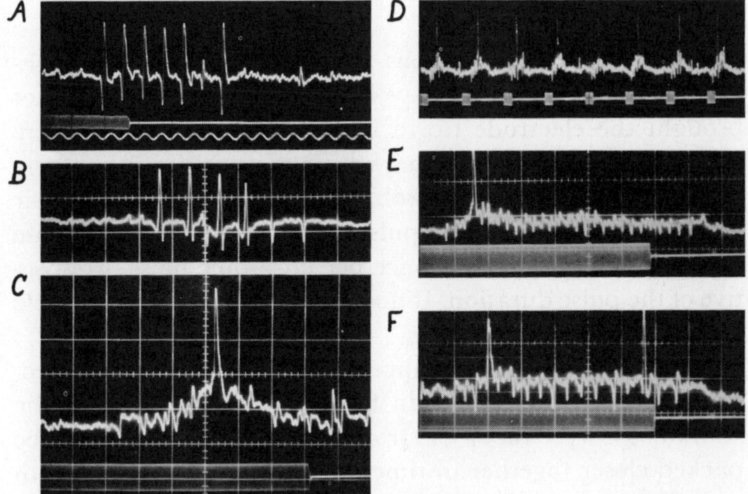

Fig. 53. Records made with a microelectrode from the neuropile of the mesothoracic ganglion. Upper trace, from microelectrode; stimulus (lower or middle trace) begins at left. (*A*) *Caenurgina erechtea,* spikes in *A* axons produced by a single 50-kc/s pulse at +8 db. Time marker, 1000 c/s. (*B*) *A* spikes (small deflections) followed by series of repeater spikes (large deflections) in response to a similar stimulus. Vertical lines mark 2.0-msec intervals. (*C*) *Heliothis zea,* response of pulse-marker interneuron to a 40-kc/s pulse +20 db in intensity. *A* spikes are small deflections preceding the single pulse-marker spike. Verticals mark 2.0-msec intervals. (*D*) The same species, response of *A* spikes (small deflections) and pulse-marker (large deflections) to a series of 5-msec pulses repeated 40/sec. (*E*) The same species, response of *A* spikes (small deflections) and pulse-marker (large deflection) to a single 35-msec pulse at +20 db. Verticals mark 5.0-msec intervals. (*F*) The same preparation and stimulus. Stimulus intensity has been reduced to +5 db and two pulse-marker spikes occur.[44]

rapidly than the *A* neurons, and apparently serve to relay the signal from the ear relatively unchanged to the opposite side of the ganglion and up to the brain of the moth. However, since they produce only minor transformations in the incoming signal, the repeaters are of little interest in our present search.

Less frequently encountered but much more interesting

were the *pulse-marker* neurons. The full significance of pulse-marker responses would have been missed if chance had not brought the electrode tip to places in the neuropile where spikes from *A* and pulse-marker units were simultaneously evident. These sites were probably close to the point of their synaptic interaction. The pulse-marker is so called because typically it responds only once per ultrasonic pulse, irrespective of the pulse duration. It does this with a long and variable latency.

In Fig. 53*C* the *A* spikes appear as small down-going deflections. They recur in a train for about the duration of the stimulus (lower trace). A stronger pulse causes them to be packed closer together in time, and the first *A* appears with a shorter latency. This could have been predicted from the experiments described in Chapter 4. The pulse-marker response is signaled by a large upward deflection occurring only after several *A* spikes have appeared. It occurs sooner with a stronger stimulus. A wide range of stimulus strengths brings about a single pulse-marker spike irrespective of the duration of the ultrasonic pulse. This is not to be ascribed to sluggish behavior, for the pulse-marker can be induced to fire by ultrasonic pulses as short as 1 millisecond in duration and it will follow each short ultrasonic pulse at repetition rates up to 40 per second (Fig. 53*D*).

One is immediately curious about the significance of the pulse-marker's behavior and about the manner in which it is produced by the train of *A* impulses. The first question will be considered later. Some of the conditions for pulse-marker activation were determined by bombarding the ear with different stimulus patterns and intensities. It was noticed that very weak stimulus pulses some tens of milliseconds in duration produced a ragged and erratic train of *A* spikes. Such a stimulus would often elicit more than one pulse-marker spike, while stronger stimuli failed to do so (Fig. 53*E*, F). This observation when coupled with closer inspection of the *A* spike pattern preceding the pulse-marker spike suggests the following conditions for pulse-marker excitation. The pulse-marker fires through temporal summation when three or four *A* spikes impinge upon it with sufficiently short interspike intervals. But it discharges then only if the group of *A* spikes is preceded by a few milliseconds of *A* inactivity. Silence apparently cocks the pulse-marker so that it is able to respond to an *A* train. This recocking cannot occur when a moderately strong and long sound pulse causes a long train of closely spaced *A* impulses. Multiple responses of the pulse-marker to a long but very weak stimulus are explained by the fact that the *A* spikes generated by such a stimulus recur in ragged groups separated by intervals of varying length. These inactive intervals are now and then sufficiently long to recock the pulse-marker, which is then fired by the next group of *A* spikes having suitable temporal proximity.

Other types of interneuron response were found by probing the neuropile. The following were met with relatively infrequently, but in each instance they were held at the electrode tip for sufficient time to establish some of their properties. They are mentioned here because when taken together they furnish some idea of how the neural machinery of turning-away is organized.

One of these interneurons has been called the *train-marker*. It is inactive during acoustic silence, but when the ear is exposed to ultrasonic pulses recurring at, say, 20 per second, it begins to fire off at some frequency of its own, perhaps 200–300 per second. This frequency is steadily maintained during the intervals between pulses and continues throughout the pulse train, ceasing only when stimulation is discontinued. Thus, the activity of this interneuron marks the duration of the ultrasonic pulse train, but from its signal one cannot tell the length of the individual pulses or of the intervals separating them.

Another type of interneuron is inhibited during and for some time after the arrival of each train of *A* impulses (Fig. 54*A*, B). It has been encountered only once, but during that experiment the electrode tip was kept near it for three hours, so that its response could be carefully studied. The site yielded small spikes from repeater neurons stimulated by *A* impulses from the ear on the opposite side. The inhibited neuron fired regularly at about 90 spikes per second, only ceasing during the arrival of the repeater train. This interesting unit is probably part of the neural mechanism concerned in turning, but since it was only met with once there is little more that can be said about it.

The interneurons described so far were found in the neuropile on either side of the mesothoracic ganglion, and seem to receive their main synaptic drive from either one ear

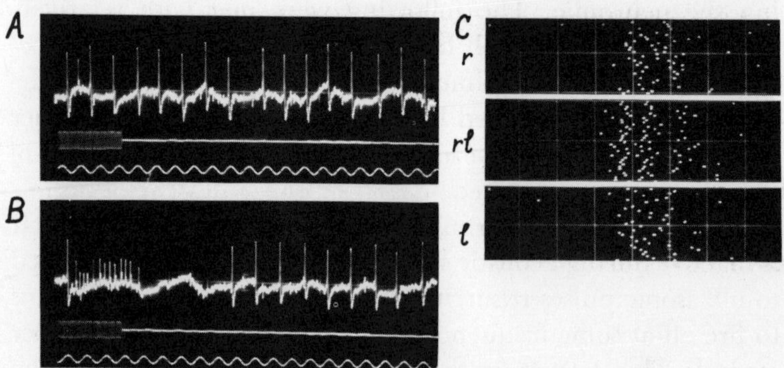

Fig. 54. (*A*) *Caenurgina erechtea,* crossed and inhibited interneuron. Stimulus (middle trace) applied to ipsilateral ear has no effect. Time marker, 100 c/s. (*B*) The same, stimulus pulse (40 kc/s at +20 db) applied to contralateral ear. A train of crossed repeater spikes is accompanied and followed by temporary suppression of the large spike.[44]

or the other. Probing in the midline of the ganglion occasionally located an interneuron that accepts and sums the nerve signals coming from both ears. This summation was especially noticeable at low intensities of acoustic stimulation. It is more evident when the responses to a number of consecutive acoustic stimuli are surveyed simultaneously. This has been made possible in Fig. 54*C* by converting each spike from the bilaterally summating interneuron into a dot. Each horizontal row of dots registers the neuron's response to a single ultrasonic pulse, the response to the next pulse being laid down directly beneath that to the first, and so on. Thus, the response to 30 pulses reads like 30 lines of print on a page, and the total response to a given mode of stimulation can be measured by adding up the number of dots. When both ears are stimulated at a fixed intensity, the neuron shown in Fig. 54*B* delivered about twice the number of spikes (dots) for a given number of ultrasonic pulses as when it was stimulated by impulses coming from either ear alone.

What sense is to be made of this grab-bag of interneurons? Can a picture be constructed when only a few randomly selected pieces of the puzzle are at hand? The pieces were assembled through chance probing and there are few indications of their order or relations to one another. My object is to understand the neural mechanism that accepts the right and left *A* receptor signals, extracts the difference between them, and applies this difference to the direct flight muscles. Stated another way, I hope to understand the sequence of transformations that link an ultrasonic pulse 0.01 second long to a change in wind direction lasting 2 seconds or more (Fig. 50*C*). I know that I am still far from this understanding, but, unless the evidence at hand leads me to some *ad hoc* hypothesis, there is little incentive to plan new experiments. The following ideas must be accepted as just that.

One encouraging aspect of the interaction between bats

and moths is that the natural signal—the series of chirps emitted by the predator (Fig. 6)—is fairly easy to mimic by electronic means, at least insofar as a moth's ear is concerned. This means that one can vary at will various parameters or dimensions of the stimulus and determine the effect of these variations on a moth's responses. These parameters are diagrammed in Fig. 55 and listed below.

(1) The pulse frequency or pitch. The pitch of most bat cries and the sensitivity of most moth ears lie roughly between 15 and 80 kc/s. A frequency of 40 kc/s was used in most of the ganglion-probing experiments.

(2) The amplitude or intensity of each pulse. Bat cries range in intensity up to 10 dynes/cm^2 and possibly much more at a distance of 1 meter.[13] Moth ears detect intensities as low as 0.02 dynes/cm^2. The range of a moth ear has been discussed in Chapter 6.

(3) The duration of each pulse. Bat cries commonly range in duration from 15 milliseconds down to a millisecond or so. In the probing experiments 10-millisecond pulses were routinely used, but wide variations were tested.

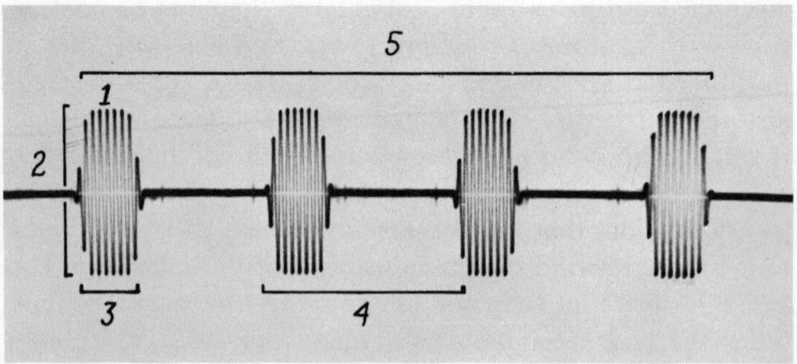

Fig. 55. The parameters of an ultrasonic pulse train. (*1*) frequency; (*2*) intensity; (*3*) pulse duration; (*4*) pulse repetition rate or interpulse interval; (*5*) pulse train duration.

(4) The pulse repetition rate. Bats commonly pulse at 8 to 10 per second when cruising in open territory, but the rate may rise to over 100 per second if the bat detects a source of echoes. Various repetition rates were also tested in the probing experiments.

(5) The pulse train length. Bats pulse continuously while flying in the dark. From a moth's viewpoint the pulse train length is determined by how long a bat remains within earshot.

These five parameters will now be examined in relation to the performance of the *A* cells and some of the interneurons described above in order to see how the ultrasonic signal might be processed by the moth's nervous system.

The message transmitted by the *A* receptor cells has been described in Chapter 4 and is depicted in Figs. 9, 10, and 53. It was shown that moths appear to be tone deaf. This means that the first parameter is left out of the *A* spike pattern. As long as the pulse frequency lies somewhere within the ear's range an observer stationed on the tympanic nerve could not determine what that frequency was. Pulse intensity could be inferred from the interval between spikes—the closeness of their spacing, from whether one or both *A* receptors responded, and, when the response was compared with that coming from the opposite ear, from the difference in response time (page 47). Pulse duration could be derived approximately from the duration of the spike train, although this might be slightly ambiguous with pulse intensity. The fourth and fifth parameters, pulse repetition rate and pulse train length, could be accurately read from the signal transmitted by the *A* axons.

Farther downstream, a single pulse-marker neuron omits from the signal it generates any measure of the intensity or the duration of the ultrasonic pulses reaching the ear. At any rate, this is true for a wide range of stimulus strengths. Relative intensity of the signal reaching right and left ears could be determined by an observer in a position to compare the departure times of impulses in right and left pulse-markers, for the pulse-marker response has a markedly longer latency with low intensities. But, since a single pulse-marker delivers only a single pulse at all pulse durations of moderate intensity, this parameter could not be measured from its signal. Parameters 4 and 5, pulse repetition rate and pulse train duration, could be clearly read from a pulse-marker response.

The response of the train-marker represents a further stage in this transformation of the stimulus. It fires steadily at its own frequency during the time that the sequence of pulses is reaching the ear. In addition to omitting the first three parameters its signals also do not mark individual pulses, This means that they indicate only the fifth parameter of the original stimulus—the duration of the pulse train. It is worth noting that the sustained changes in wing angle that occur during an attempt made by a moth to turn away (Fig. 51) are probably produced by a similar pattern of motor nerve impulses transmitted to the appropriate direct flight muscles. However, no direct relation between these two actions has yet been established.

This picture of central transformations does not take into account the behavior of the bilaterally summating and the inhibiting interneurons. Furthermore, it completely lacks anatomical substance. The neurons and their connections have not been seen in the flesh; only their voices have been heard. Also it is not known in what order the interneurons are activated or how the differential effect on the right and left wings is accomplished. However, the picture does suggest how the various parameters of the ultrasonic pulse pattern might be transformed into the behavioral response.

Because they have been progressively eliminated during the neural processing of the stimulus it must not be thought that these parameters are superfluous. Each may be likened to a key that becomes useless only when the lock that it fits has been opened. The moth reacts optimally only to sounds having certain frequencies, intensities, pulse repetition rates, and pulse durations. The interneurons so far revealed by probing with the microelectrode may be part of the system that checks or passes on these parameters, each unit transmitting its signal to the next stage if a given parameter lies within certain limits.

At this point we have reached the soft ground to be found at the edge of every area of research. Electrophysiological methods of tracking messages in single nerve fibers require one to follow narrow paths through the central nervous system and to disregard all irrelevant signals passing back and forth in neighboring threads of the neural mesh. They force one to regard the system with tunnel vision and must be supplemented by a wider view. What happens to behavior when whole sections of neuropile are removed and the multifarious transactions within are silenced? Surgery on the insect nervous system is a much older art; its results will be examined in the following chapter.

12. *The Insect Brain*

In Chapter 2, I compared the activities of psychologists, ethologists, physiologists, and others searching for the mechanisms of behavior to those of workers tunneling into a mountain from many directions and with many kinds of equipment and many tactical objectives. The chapters that followed dealt mainly with a single category of this multifarious activity—reconstruction of certain behavioral patterns of insects from what is known about the properties of neural components such as acoustic receptors in moths, synaptic processes, and the like. The properties of the neural mechanisms were our primary concern and point of departure, so this may be called the "mechanisms" approach. But a wide gap remains between the mechanisms and the behavior; in other words, the hole that we have dug in the mountain is both small and shallow, and our main comfort must be that we have not yet lost our bearings with the surface realities of basic morphology and physiology.

Another method that contrasts sharply with the mechanisms approach to these problems is the "systems" approach. Unlike the former, the systems approach ranges far, fast, and wide. Its power and its weakness reside in the fact that it is not limited by nor concerned with mechanisms, for its hypotheses and conclusions can be displayed equally well by models built from electrical or mechanical mechanisms, or merely by mathematical symbols. When the systems approach can be connected to the mechanisms approach so that its feedback loops and automata become clothed in flesh and blood, we shall see real and exciting progress in understanding behavior.

The present chapter presents a few experiments that are primarily concerned with the systems of insect behavior. Included also are some of the efforts to connect the systems with mechanisms.

The bulk of our information on the operation of the insect nervous system derives from studies in which parts of the nervous system are removed by surgery, and their function inferred from the behavior of the rest of the animal (see Fig. 1*E*). Classical experiments of this kind began with Aristotle, who records observations on the behavior and responses of the isolated body segments of a wasp. The nineteenth and early twentieth centuries were times of great activity, associated with names such as Bethe, von Uexküll, Kopec, von Buddenbrock, Baldi, Baldus, Alverdes, and Ten Cate. The review [56] by the last named is an excellent summary of this classic period.

The advantages of the ablative or surgical method are that its minimum requirements are sharp eyes, steady hands, and a pair of ultrafine scissors. Its disadvantages are that the direction of the cuts made in the nervous system bear only the crudest relation, and often none at all, to the multitude of functional nerve pathways that are being separated or interfered with. When a telephone cable containing hundreds of lines is accidentally severed, an outside observer cannot learn much more from the effects of this misadventure than that the cable supplied a certain area of the country. The actual losses in business, personal misunderstandings, and other disjunctions in the community do not come to light except to those personally concerned. However, the method of ablation has supplied a large amount of information about the interaction of different parts of the insect nervous system—a fact that must be attributed to the high degree to which some specific nerve functions are localized in various parts of the insect brain and nerve cord.

Modern electronic means of producing highly localized stimulation and of destroying and marking very small and circumscribed areas of nerve tissue have enabled Huber[19] and Vowles[62] to apply the method of ablation with a precision that begins to match the electrophysiological method. The latter, comparable to eavesdropping on single conversations in a multichannel telephone cable, has limitations of its own that have been mentioned a number of times in the preceding pages. This promise of a closer approach between the adherents of surgery and those of electrodes—the "hackers" and the "pokers"—makes it worth while to compare the generalizations about insect behavior that have been arrived at by both.

The classical ablation experiments mentioned above showed clearly that each ganglion of the ventral nerve cord is autonomous for a number of local coordinated movements and reflexes such as stepping and grasping in the leg-bearing segments. Coordinated intersegmental movement of the legs in walking is dependent upon connections between the segmental ganglia, and the whole operation is turned on and off and steered by nerve centers in the head ganglia.

Walking and Turning in the Mantis. The possibilities and limitations of ablation are illustrated by experiments on the praying mantis.[33,36] The form of its central nervous system is shown in Fig. 56. Mantids are by nature inactive creatures, perching for hours or days on a twig or flower and striking at passing insects from the ambush provided by their immobility and sticklike or leaflike appearance. If disturbed from its perch a mantis is able to run with some speed, but it rarely travels far. If a mantis is placed on the ground it wanders rest-

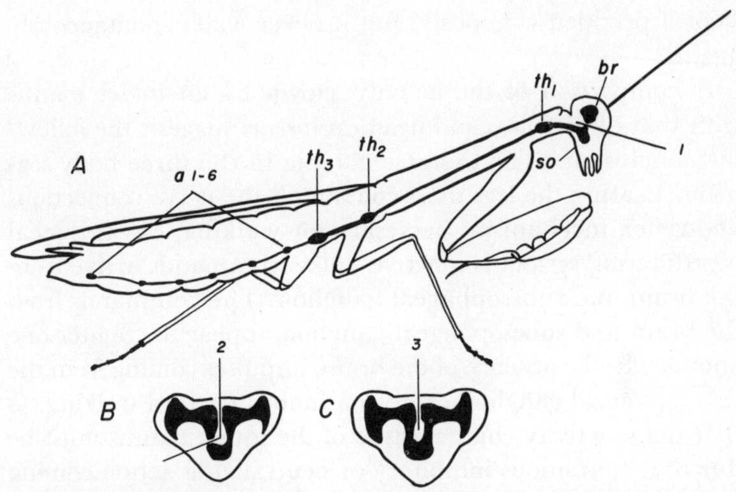

Fig. 56. (*A*) Diagram of the central nervous system of the praying mantis. *a 1-6,* abdominal ganglia; *br,* brain; *so,* subesophageal ganglion; *th 1-3,* thoracic ganglia. (*B, C*) Anterior views of brain and subesophageal ganglion. *1,* cut to produce brainless mantis; *2,* half-brained; *3,* split-brained.

lessly until a vertical surface is encountered. This it climbs immediately, and resumes its ambush at or near the top.

Removal of the eyes or antennae does little to alter this pattern of activity, although, of course, the former operation eliminates the possibility of visual orientation and insect capture. Surgical removal of the supraesophageal ganglion, or brain (Fig. 56*A*), causes the insect to walk continuously. One is reminded of a driverless car in forward gear, for the mantis is unable to climb, turn, or avoid obstacles, and eventually becomes firmly wedged in some obstruction. Removal of the subesophageal ganglion in addition to the brain (equivalent to decapitation) eliminates all spontaneous walking except for the continuous rotary movement seen only in mature males (Chapter 10). The headless insect stands, and rights itself when upset, and it can also be made to take a few coordinated steps if prodded vigorously. But it never walks spontaneously again.

A comparison of the activity shown by an intact mantis with that of brainless and headless insects suggests the following conclusions. The thoracic ganglia in the three body segments bearing the legs must contain all the nerve connections and reflex mechanisms necessary for walking, but this local coordinating system is inactive unless commands arrive from the brain and subesophageal ganglion. The commands from the brain and subesophageal ganglion appear to counter one another. In the absence of the brain, impulses coming from the subesophageal ganglion cause continuous forward walking, so that the inactivity characteristic of the intact mantis must be due to a continuous inhibitory or neutralizing action coming from the brain and suppressing the excitatory action of the subesophageal ganglion.

Two more surgical experiments add to the picture. Removal of one half of the brain (right or left), leaving the other half connected to the nerve cord (Fig. 56*B*), causes the mantis to turn restlessly in a tight circle toward the intact side. This turning is quite unlike the rotary locomotion seen in decapitated males. It has all the characteristics of the sharp turn often made during normal walking. The head and body bend in the direction of the turn, while the legs walk forward on the "brainless" side and backward on the intact side. It was once thought that following removal of the brain on one side the insect merely pivoted because the legs were more active on that side of the body lacking a brain, but a closer look at the subject shows that all the legs participate actively in the turning, each being lifted and placed in perfect coordination with the others but in a specific direction of its own. The turning movement of a half-brained mantis appears to be identical with that made by an intact insect while walking, except that it never ends.

The final experiment consists of making a cut between the right and left halves of the brain so as to separate them from each other while leaving each connected to the nerve cord (Fig. 56*C*). For several months I was greatly puzzled by the results of this operation, because the split-brained mantis appears to be completely normal under some circumstances. If observed while hanging from a natural perch such as a twig or the ceiling of the cage, it shows vigorous visual reactions to movements made nearby, and is able to catch flies and feed. It shows no tendency to move to another spot, but neither does an intact mantis under the same circumstances. The nature of the deficiency becomes obvious only when the mantis is placed on a table top or on the floor of its cage. An intact mantis would immediately walk to the nearest vertical object and climb to the top, whereas the split-brained mantis makes continual reaching movements with its front legs toward the nearest twig, but seems to be incapable of walking toward it. All its movements are of the nature of intention movements —at any rate they suggest to the observer familiar with the normal behavior of mantids that walking is about to take place —but not a step forward is taken. At the same time the head is turned repeatedly and the front legs are continually reaching out. If one attempts to prod the split-brained mantis into walking, it may take a few steps backward, but it cannot be made to walk forward.

The behavior of the half-brained and split-brained mantids considered in relation to that of the brainless and headless insects suggests two additional conclusions. The first is that the inhibition imposed by the brain on forward locomotion could be described as a right and a left *turning command* imposed from corresponding halves of the brain on a *forward command* transmitted from the subesophageal ganglion to the thoracic motor centers. Movements of the intact insect are visualized as being determined by changes in the relative potencies of forward command, right turning command, and left turning command acting on the thoracic centers. The second conclusion is that the systems generating the right and left turning commands mutually inhibit one another through the mid-line connection between the brain halves. Thus, an increase in the right turning command partially suppresses the left turning command and vice versa, both effects being always less than if they were acting without the inhibitory influence of each other. At the same time, both turning commands act by suppressing the forward command.

In the split-brained mantis mutual inhibition by the turning-command systems has been eliminated so that both turning tendencies become enhanced to the point where they completely suppress the forward command generated in the subesophageal ganglion, and forward movement is blocked. The enhanced turning tendencies are manifest in the raised posture due to increased muscle tonus on both sides of the body, and by the very frequent intention movements that suggest to the observer that turning is about to occur. However, turning is never completed, partly because both right and left turning tendencies are enhanced to the same degree and tend to cancel one another, and partly because they work together in blocking the forward tendency generated by the subesophageal ganglion. In the half-brained mantis continuous turning toward the intact side would then be due not only to the elimination of the opposite turning tendency brought about by removal of the opposite half brain, but also to the absence of the inhibition to turning that is normally supplied by the missing brain region. This interpretation of the experiments is shown diagrammatically in Fig. 57.

This story illustrates both the strengths and the weaknesses of the ablation method. Concepts of the sort described above are heuristic, that is, they suggest further experiments based on what is known about the properties of neural mechanisms.

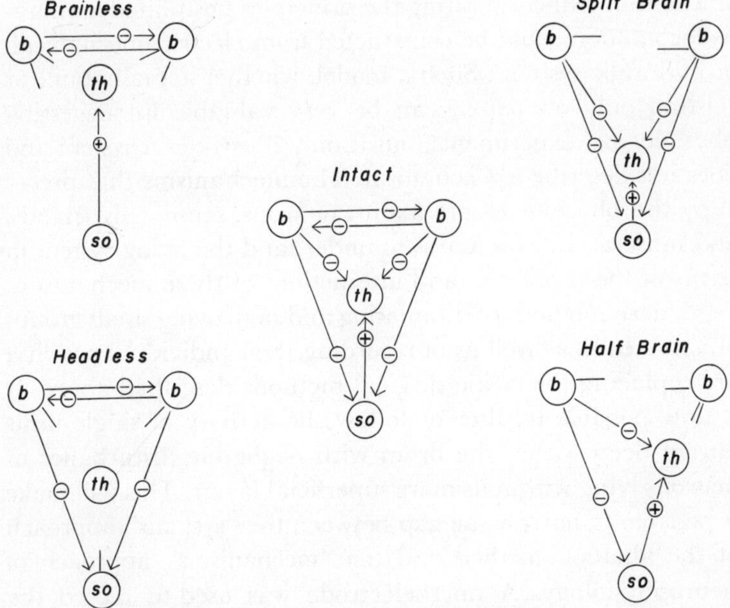

Fig. 57. Representation of hypothesis regarding the effects of brain operations on locomotion in the praying mantis. *b,* brain; *so,* subesophageal ganglion; *th,* thoracic motor complex. Arrows show the direction of action (*plus,* excitation; *minus,* inhibition).

For instance, it would be worth while to find out whether the source of the forward command takes the form of endogenous impulses arising in neurons within the subesophageal ganglion and traveling in axons within the nerve cord to the thoracic motor centers. One might expect that this nerve activity would increase when the brain is removed. The experiment might be planned along the lines of those described in Chapters 9 and 10. However, it is important to note that the ablation method cannot answer questions about mechanisms, and terms like "turning tendency" and "forward command" have no physiological meaning. This is illustrated by the fact that a model or automaton incorporating the principles postulated for mantis locomotion could be constructed from electric, mechanical, or hydraulic systems. Such a model, whether actually built or existing only on paper, can be very valuable in suggesting physiological experiments, but it only illustrates a system and does not describe a mechanism. The mechanisms that preoccupy the physiologist are axons, neurons, sense cells, glands, and muscles, and his aim is to understand the living system in terms of the properties and interactions of these mechanisms.

Modern methods of stimulating and destroying small groups of nerve cells, as well as of recording their individual activity, are replacing the crude surgical methods described above. It is now possible to alter or follow the activity of single units buried deep within the brain with negligible disturbance of neurons lying within its more superficial layers. This will make it possible to narrow the gap between the "systems" approach of the ablation method and the "mechanisms" approach of neurophysiology. A microelectrode was used to record the electrical responses of interneurons in the thoracic ganglia of a moth when its ears were being stimulated with ultrasonic pulses (Chapter 11). A similar electrode can be used to inject current, that is, to stimulate small groups of neurons lying deep within the central nervous system while changes in the behavior of the insect are being observed.

The most detailed study is that being carried out by Huber[19] on the brain of the cricket. He inserts a single or double electrode 10 or 30 microns in diameter and insulated up to its tip into the cricket's brain, and cements it in place in such a manner that it does not interfere with the subject's behavior (Fig. 58). The cricket is kept in one place for close observation by a support cemented to its back, but it is able to walk freely on a cork ball held in its feet. The rolling of the ball gives a measure of the speed and direction of locomotion. This restriction does not appear to hamper unduly the behavior of crickets and most other insects, and it has the advantage

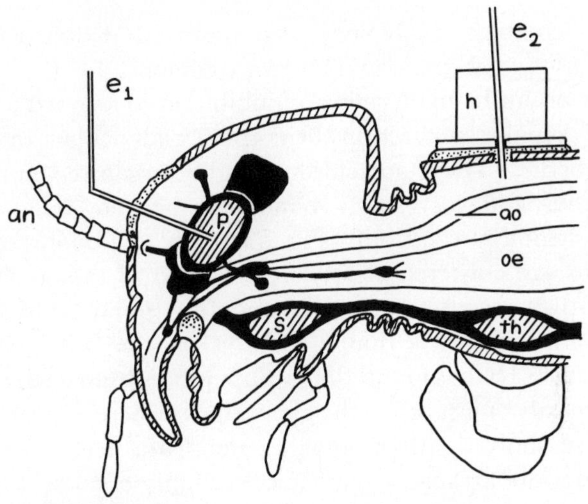

Fig. 58. Longitudinal section through the head and prothorax of a cricket. Structures cut by the section are hatched. *an,* antenna; *ao,* aorta; e_1, insulated electrode inserted into the brain through an opening in the cranium and held in place by the blood clot in the wound; e_2, indifferent electrode connected to holder *h* and inserted into the thorax; *oe,* esophagus; *s,* subesophageal ganglion; *th,* first thoracic ganglion. (Modified from Huber.[19])

that walking, grooming, flying, singing, mating, and other activities can be observed closely.

This arrangement makes it possible to stimulate nerve tissue adjacent to the electrode tip with electric shocks of various intensities and frequencies while observing the behavior of the unanesthetized cricket as it stands on the cork ball. At the end of the observation period high-frequency alternating current is passed through the electrode. This treatment kills and coagulates a patch of nerve tissue around the electrode tip, making it possible to observe the effects on behavior of the absence of the nerve tissue that had previously been stimulated. At the same time the exact position of the electrode tip is marked by the coagulum, and can be determined by microscopic examination *post mortem.*

This combination of local stimulation and destruction of nerve tissue has enabled Huber to demonstrate that in the cricket brain the main source of inhibition to forward locomotion is actually localized in the corpora pedunculata or mushroom bodies. These are curious, highly organized groups of numerous and very small monopolar neurons lying in the dorsal region of the brain (Fig. 59). The cell bodies of these neurons are clustered closely in the head of the mushroom.

Another recent anatomical and functional study of the mushroom bodies is that of Vowles on the brains of bees and ants.[62] He has found that a fiber passes from each neuron into the calyx (just below the glomerulus) where it gives off one or more short tightly branching dendritic processes. These dendrites are arranged in a complex highly organized fashion within the calyx, where they appear to receive synaptic contacts from axons entering the calyx from other parts of the brain. The fibers from the mushroom-body neurons then continue as an ordered bundle, each dividing once more into the fiber bundles known as the alpha and beta lobes. Vowles concludes that the mushroom-body neurons are internuncial cells lying entirely within the mushroom bodies, and that they receive in the region of the calyx synaptic contacts from fibers originating in the optic and antennal ganglia, forming in their turn synapses within the beta lobe with fibers passing to the motor systems of the subesophageal ganglion and nerve cord. In the alpha lobe the mushroom-body neurons make synaptic connections with the antennal and visual systems, but it has not been clearly established whether those connections are afferent or efferent. Vowles considers this arrangement (Fig. 60) of two widely separated dendritic regions on each mushroom-body neuron to be a significant feature connected with the integrating function of the mushroom bodies.

The mushroom bodies have long been suspected of being "higher centers" in insects, although there has been little evi-

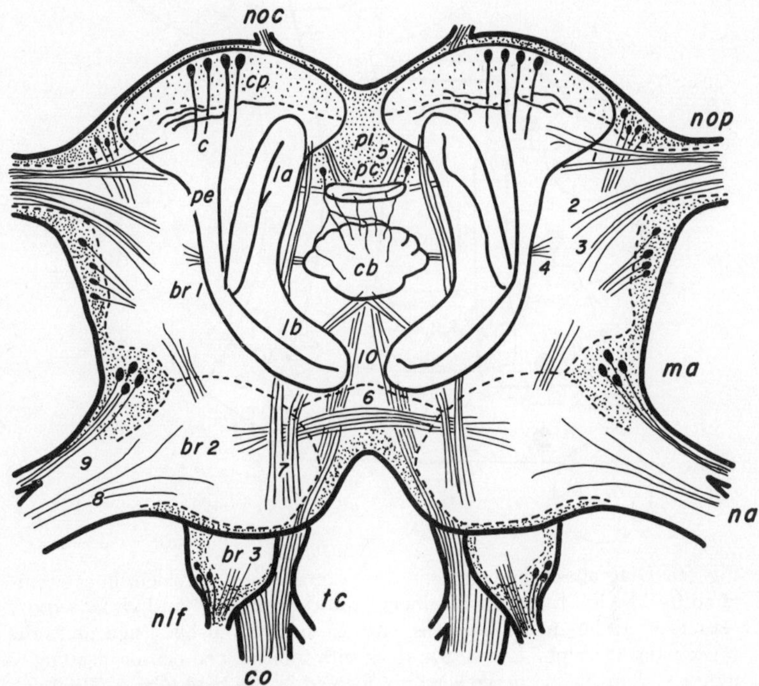

Fig. 59. Diagram of frontal section through the brain of a cricket. *Stippled regions* are occupied by nerve-cell bodies; *clear regions,* neuropile. *br 1,* protocerebrum; *br 2,* deutocerebrum; *br 3,* tritocerebrum; *c,* calyx; *cb,* central body; *co,* circumesophageal commissure; *cp,* corpus pedunculatum or mushroom body; *la,* alpha lobe; *lb,* beta lobe; *ma,* antennal motor center; *na,* antennal nerve; *nlf,* labrofrontal nerve; *noc,* ocellar nerve; *nop,* optic tract; *pc,* pons cerebralis; *pe,* pedunculus or stalk of mushroom body; *pi,* pars intercerebralis; *tc,* tritocerebral commissure. Tracts within the brain: *1, 2, 3,* branches of the optic tract, one branch entering the calyx; *4,* optic commissure; *5,* ocellar tract; *6,* antennal commissure; *7,* olfactorio-globularis tract, one branch passing from the antennal lobe to the calyx; *8, 9,* sensory and motor roots of antennal nerve; *10,* tracts connecting central body with ventral nerve cord. (Modified from Huber.[19])

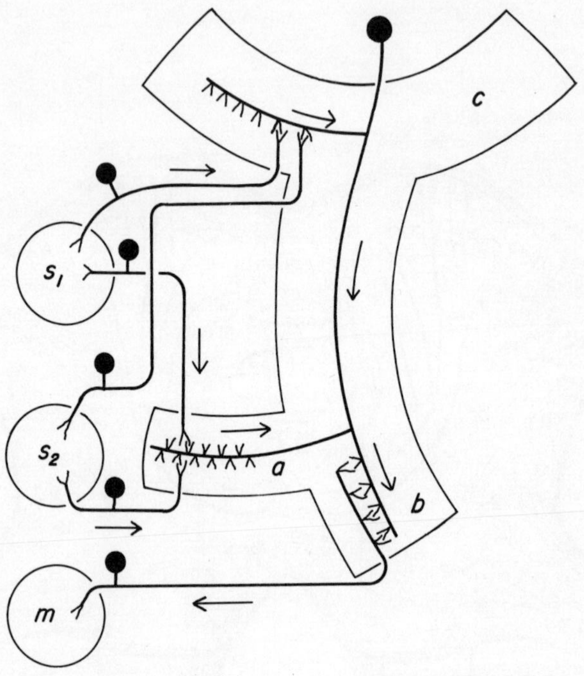

Fig. 60. Diagram of the connections of a mushroom-body neuron as postulated by Vowles. Internuncial fibers from the antennal and visual sensory centers S_1 and S_2 make synapses with the mushroom-body neuron in the calyx *c* and the alpha lobe *a*. Synapses with internuncial neurons leading to motor systems in the nerve cord are formed in the beta lobe *b*. (Redrawn from Vowles.[62])

dence of this beyond their orderly complexity. They are especially well developed in the worker castes of honeybees and ants, and Vowles has shown that in ants disruption of the mushroom-body connections with the optic centers interferes with learned patterns of maze running, without, however, eliminating all reactions to visual stimuli. Now Huber has shown that the mushroom bodies are sources of inhibition to locomotion, and that, as in the praying mantis, the right and left inhibitory systems exert a mutual inhibitory effect upon one another.

In crickets the subesophageal ganglion appears to play a part in the excitatory control of behavior, but this role is much less marked than in mantids. Huber finds that the excitatory system in crickets depends upon combined effects of the subesophageal ganglion and the central body (Fig. 61). This is a highly organized region of densely intermingled neuron terminations lying between and below the mushroom bodies. Like the latter, it appears to receive many fibers from the optic and antennal ganglia. Huber concludes that the central body and subesophageal ganglion act in concert in exciting the locomotor mechanisms in the thorax, and that both are in turn regulated through inhibition originating in the mushroom bodies (Fig. 61).

Turning to either the right or the left can be elicited by stimulating a variety of sites in either half of the cricket brain. This suggests that the neural mechanisms steering locomotion are highly complex and are not confined to a specific center. We must add this example to the growing list of experiments in which study of brain function at a finer and more precise level has forced us to discard the comforting concept of spatially localized functional nerve centers. The mechanisms of right and left turning commands deduced from the mantis experiments are probably based upon a three-dimensional brain-wide network of connections that is somehow superimposed upon similar networks controlling other activities. Stimulation or destruction at any of many widely separated points in such a network appears to alter its general effectiveness without causing specific defects. Thus, the balance between the right and left turning commands is probably disturbed by almost any asymmetrical stimulation or injury to the brain. One is understandably reluctant to accept this field concept in place of the center concept because it is not only difficult

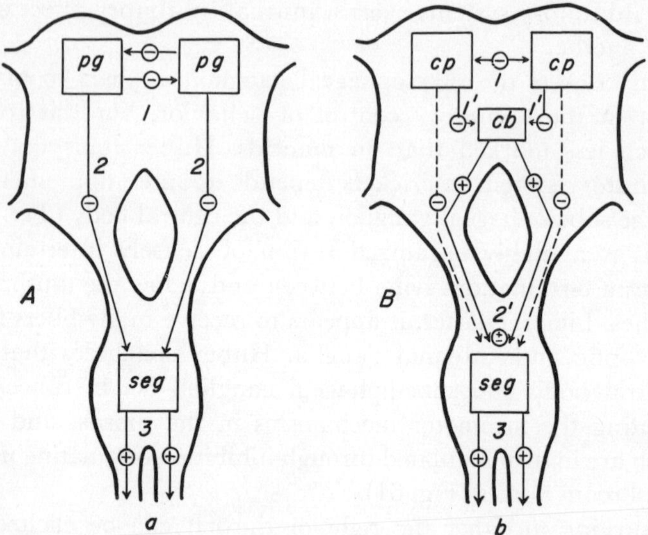

Fig. 61. Scheme showing interaction of the brain and subesophageal ganglion in controlling locomotor activity in (*A*) the mantis and (*B*) the cricket. *cb,* central body; *cp,* corpora pedunculata or mushroom bodies; *pg,* protocerebral ganglion; *seg,* subesophageal ganglion; *1,* mutual crossed inhibition between protocerebral ganglia or mushroom bodies; *1′,* inhibition of central body by mushroom bodies; *2,* inhibition of subesophageal ganglion by protocerebral ganglia; *2′,* steering of the activity of the subesophageal ganglion by excitatory and inhibitory action from the mushroom bodies and central body; *3,* control of the motor centers in the thorax. (From Huber.[19])

to visualize in terms of neural mechanism but harder still to dissect experimentally. Yet many experiments on other animals and behaviors, such as those of Lashley[21] on learning in rats and of von Holst and von Saint Paul[17] on innate behavior patterns in chickens, leave no alternative.

Huber[19] has also shown that the mushroom bodies and central body of the cricket brain constitute the major control system for song production. Male crickets sing by rubbing a file on one elytron (forewing) over a scraper on the other. Their repertoire commonly consists of three types of song.

The *calling song* is produced by a male in mating condition, and serves to attract females. If another male is encountered, the *rival song* takes its place, and is usually a prelude to territorial combat. If a female is encountered, the male replaces the calling song by the *courtship song*. The three songs of a given species differ in their construction mainly in the pulse pattern, determined by the intervals between and the duration of the separate elytral movements, although there are also differences in the pitch and intensity of the individual pulses.

The second thoracic ganglion, as segmental center for the elytra, contains all the coordinating mechanisms for operating the muscles concerned in song production. No spontaneous singing occurs if the brain is removed, although a normal performance is possible if only one mushroom body is present. Huber found that stimulation of specific areas in one mushroom body would release in a reproducible manner each one of the three song types, or would inhibit spontaneous singing already in progress. Stimulation of the central body produced only an atypical song pattern, often after a shorter latency. Activity in specific sensory pathways connected with localized regions of the mushroom bodies appears to determine selection of the song type. This information, acting through the central body, is transmitted to the thoracic centers. Huber also points out that the elytral movements made by singing crickets probably evolved as a modification of the wing movements made by flying insects. It would be worth while to investigate the role of the mushroom bodies in connection with flight in other insects, since this activity is always associated with a suppression of walking movements.

The Strike of the Mantis. It is impossible to leave this type of approach to the analysis of insect nerve mechanisms without mention of the elegant experiments of Mittelstaedt[27] on prey capture in the praying mantis. However, unlike the experiments of Huber, which have begun to break down the system into its component neural mechanisms, the theoretical models derived by Mittelstaedt have yet to be clothed in flesh and blood. When this is done, I predict that the precision and power of Mittelstaedt's work will establish novel connections between the nerve physiology and the behavior of insects. Mittelstaedt has adequately reviewed his work in English, so I shall give only a bare outline of his story.

The praying mantis captures other insects by a rapid extension and grasping action of the forelegs (Fig. 33). The prey may be hit from a variety of angles, and the mantis is successful in 85 to 90 percent of its strikes. The muscular explosion that propels the forelegs at the prey is so rapid that it cannot be steered by the mantis during its execution, so that it appears to be a ballistic action like throwing a ball or a punch. This means that the mantis must make a prior estimate of the direction and distance of the prey relative to its own body axis, thus presetting the excitation pattern destined for the various muscle groups connected with the forelegs so that upon the strike command they contract in a pattern that will direct the movement to the prey.

Mittelstaedt has examined the systems partaking in this strike estimate by measuring the insect-catching performance of mantids under various conditions, and by using the apparatus shown in Fig. 62. The mantis, like Huber's crickets, seems to be unhindered in much of its behavior by being fixed to a rigid support by its thorax. It walks on a platform counterweighted with a weight equal to the body weight of the mantis. Its head is free to rotate about the vertical axis, head position relative to its (fixed) body axis being registered by a pointer on the horizontal scale. A living fly on a needle is presented at various angles, and the horizontal angles made by both fly and mantis head relative to the body axis are measured from the scale.

A hungry mantis follows the fly closely by making head

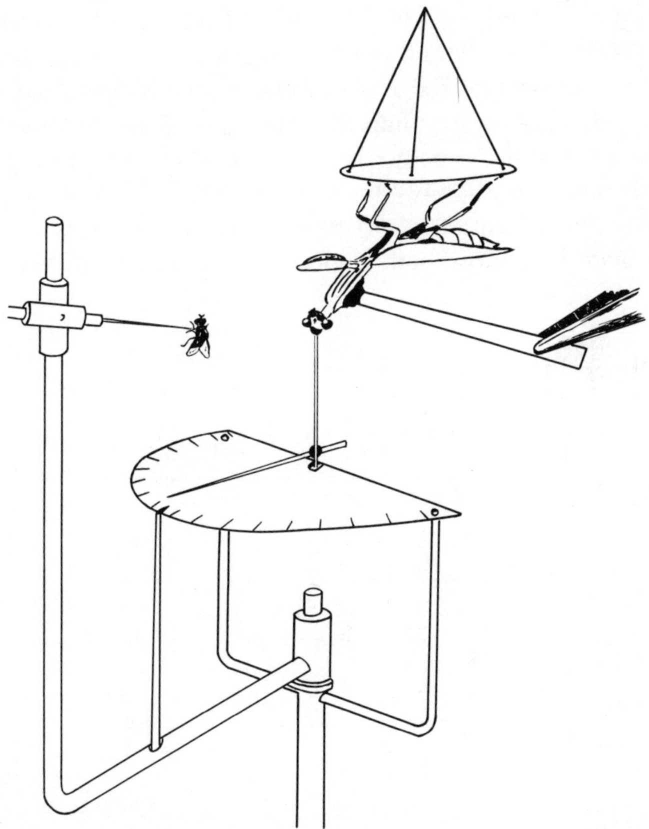

Fig. 62. Device for measuring the fixation deficit during prey localization in the praying mantis. The mantis is fixed at the prothorax. The head is free to rotate about the vertical axis, its position being shown by the balsa-wood pointer. The prey can be presented at different angles, and the angles made by head and prey measured on the same scale. (From Mittelstaedt.[27])

movements as the fly is moved back and forth in front of it. If the prey is within striking distance, the head of the mantis follows its movements smoothly and continuously; if the prey is beyond the range of the forelegs, the head of the mantis follows it in a series of saccadic jerks. Measurement at equilibrium (when both fly and head are motionless) at a number of positions shows that the angle *y* between the axes of head and body (Fig. 63*A*) is less than the angle *z* between the direction to the prey and the body axis by an amount proportional to the deviation of the prey from the body axis, at least for the smooth movements that precede the strike. Thus, when the prey is straight ahead both angles are zero and the fixation line

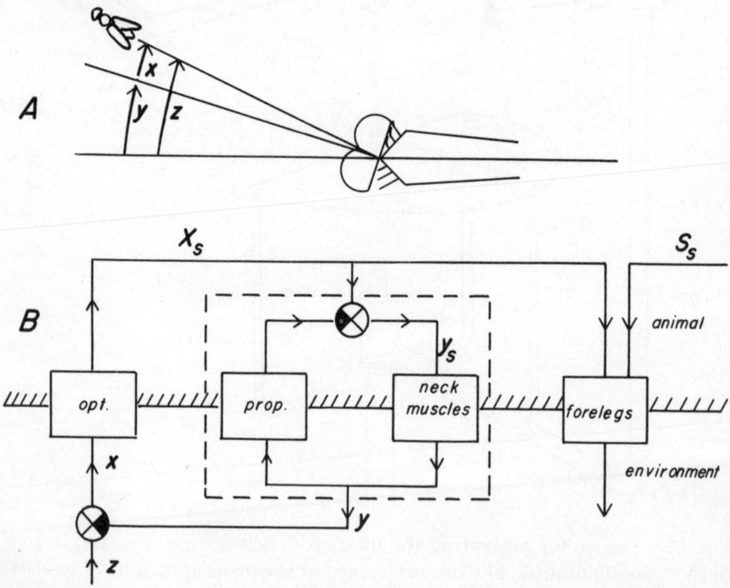

Fig. 63. (*A*) Head and prothorax of a mantis fixating on a fly: *x*, fixation deficit; *y*, deviation of head from body axis when prey is fixated; *z*, deviation of prey from body axis. (*B*) Control pattern for prey localization. Boxes represent nervous and muscular elements of each subsystem. The optic-feedback loop contains within it the proprioceptor loop (*dashed box*) that runs counter to it. The optic output x_s contains the correct aiming signal which takes effect upon arrival of the strike signal s_s. White segments within circles, indicate addition; black segments, subtraction. (Modified after Mittelstaedt.[27])

of the mantid's regard falls directly on the prey; as the prey moves to a greater angle, the line of regard lags behind the prey by a proportionally greater amount. Mittelstaedt has called this deviation of the head from the true line of regard the *fixation deficit* x. Steady-state determinations show that x has a constant value of about 15 percent of z for the continuous fixation movements that immediately precede the strike.

The system used by the mantis in steering its strike is revealed by inquiring into the origin of the fixation deficit. First, it is obvious that the head movements made in following the prey must involve an optic feedback system. Once the mantis assumes the capture "mood," asymmetry of the prey image in the binocular field of the compound eyes must generate a proportional nerve message that acts on the neck muscles so that the asymmetry is reduced. This negative feedback principle is analogous to that controlling constant length in the postural muscles of vertebrates (Chapter 8) and a host of other biological regulations. However, the asymmetry of the fly image is not reduced to zero (except when the fly is directly ahead), as one would expect in such a system working alone, but consistently fails to reach it by the amount of the fixation deficit. This suggests the presence of another constantly acting regulatory system that must operate in opposition to the optic system.

The receptors in this second system turn out to be proprioceptors in the neck. Two pairs of plates covered with small socketed hairs are found on either side at the base of the neck (Fig. 63*A*). These hair plates are arranged so that when the head is held symmetrically facing forward the back plate of the cranium bends a certain proportion of the hairs on both sides. Turning the head to the right causes greater displacement of the hairs on the right hair plates while those on the left hair plates suffer correspondingly less deflection. Sense cells at the bases of the hairs are stimulated in proportion to the deflection. Experiments in which extraneous forces, such as gravity, were applied so as to deflect the head from its "natural" position suggest that the differential of the nerve discharges from the right and left hair plates constitutes a signal that generates, through the neck muscles, a force of opposite sign to that tending to deflect the head. This is also, then, a negative-feedback system closely analogous to that maintaining constant muscle length in vertebrates (Chapter 8), serving to keep the head position constant in the presence of external perturbations. We must suppose that when the mantis freely turns its head to a new position the nerve-impulse discharge containing the turning command must exactly counterbalance the signal contained in the differential discharge from the now asymmetrically stimulated hair plates. The system continues to maintain the new position in the presence of external perturbations. In prey fixation the fixation deficit is evidence that this proprioceptive or head-straightening system is linked with a negative sign to the optic system.

Mittelstaedt measured the strike performance and errors made (right or left) after various types of interference with the proprioceptive system. Denervation of both hair plates is expected to greatly reduce or abolish the fixation deficit, so that the mantis tends to look directly at the fly irrespective of the angle of presentation. Furthermore, when the strike occurs, it should take a forward direction at all values of z, so that the mantis misses most flies except those presented directly ahead. While the evidence on the former prediction is still equivocal, the latter has been clearly verified. Denervation of the left hair plates causes the mantis to hold its head permanently 25° to 30° to the left and, consequently, to strike persistently too far to the right. Rigid fixation of the head to the left by a bridge of balsa wood cemented to the thorax causes the mantis to miss 70 to 80 percent of all flies presented. About the same percentage of strikes is directed too far to the right.

Mittelstaedt's hypothesis for the stroke-aiming system of the mantis is depicted in Fig. 63*B*. Both optic and proprioceptive feedback systems operate through the neck muscles, the proprioceptive system being represented as a loop inserted within (dashed box) and running counter to the optic loop. The nerve signal y_s acting on the neck muscles is the difference of the signals coming from the optic and proprioceptive systems, and is expressed as the head position y. The output of the optic system x_s with the head in this position corresponds to the fixation deficit x. Since it contains information about the prey-to-head angle as well as the head-to-body angle, the output of the optic system could preset the contraction pattern of the foreleg muscles so as to project them at the prey upon receipt of the strike command. This has been shown to be released visually by some slight movement of the prey.[31] Additional evidence, as well as other general solutions to this and to other control systems in insect orientation, have been fully discussed by Mittelstaedt.[27]

This interesting analysis suggests a number of additional experiments. In preparing to strike, the mantis not only turns its head, but may also turn the prothorax or even the whole body toward the prey. The regulation of these systems is probably similar. It would be worth while to work out their relations to the systems revealed by brain operations on the mantis. The part played by binocular vision in the transition from saccadic to smooth head movements as the prey comes within striking distance, as well as the accuracy with which the mantis is able to measure its own striking distance, would be worth investigating. The striking distance must increase suddenly following each nymphal moult. Some of the factors in the strike-releasing power of different kinds of prey have been investigated[31] but more remains to be done in this connection. The work also connects with the already extensive studies on insect orientation. A survey of this field is well beyond the possibilities of this book, and the reader is referred to the works of von Frisch,[12]

Lindauer,[22] Hassenstein, Jander, and others mentioned in Mittelstaedt's review.[27]

A few preliminary (unpublished) attempts have been made by Mittelstaedt and Roeder to identify and isolate the neural mechanisms postulated in the system directing the strike. Registration of spike potentials in the nerves arising from the hair plates shows that the proprioceptive message is transmitted by large numbers of small fibers, suggesting fairly precise measurement of the head deviation. The afferent-nerve discharge is proportional to head deviation, and the receptors are nonadapting, as would be required of a mechanism reporting on the tonic posture of the head. A few records were obtained from second-order fibers with which these hair-plate fibers make synapses, but we still know next to nothing about the neural mechanism of this system.

13. *Inhibition, Endogenous Activity, and Neural Parsimony*

One general impression left by these studies of control systems in the insect nervous system is that central regulation of local systems is accomplished largely through inhibition. At first glance, it would seem that equally effective control could be obtained by excitatory systems acting as triggers for local coordinating mechanisms. Since this is not the case, it is worth while to inquire into the reasons.

At a much lower level of coordination, as in regulation of the antagonist muscles of an appendage, both excitation and its antithesis would seem to be inevitable. The mechanisms concerned in this situation were discussed at length in Chapter 8. A noxious stimulus causing withdrawal of the leg not only elicits excitation and contraction of the flexor muscles, but concomitantly inhibits or blocks excitation of the more powerful extensors. This automatic system of simultaneous inhibition of antagonist muscles seems to operate in nearly all motor mechanisms. The motor neurons of an animal can be visualized as being encompassed by an excitatory network of nerve connections as well as by its negative version, the inhibitory network.

Motor Mechanisms. In the Crustacea, part or all of the motor interaction between excitation and inhibition takes place peripherally instead of within the central nervous system.[18] Crustacean muscles or whole functional muscle groups (flexors or extensors) may be supplied by no more than two to five motor axons compared with the dozens or hundreds performing the same function in vertebrates. Each nerve fiber branches to some or all of the muscle fibers in the unit, and forms endings at many points on the surface of each muscle fiber. The mode of contraction depends upon the motor nerve fiber stimulated. In some cases a rapid maximal twitch is produced; in others the contraction is slow and partial, showing facilitation that depends upon the number and frequency of the incident nerve impulses. One and sometimes two of the fibers supplying a muscle bring about no contraction when stimulated. Their action is evident only if superimposed on a contraction engendered by stimulation of the excitatory nerve fibers, when it can be seen that they suppress or block contraction. Thus, the surface of crustacean muscle fibers is a site for the integration of excitatory and inhibitory actions, performing a function analogous to that of the cell bodies of spinal motor neurons in vertebrates (Chapter 8).

The general plan of innervation of insect muscles is very similar to that found in crustaceans. A typical muscle is provided with two or three motor nerve fibers, each of which supplies from 30 to 100 percent of the muscle fibers. Hoyle[18] has shown that the large extensor muscle in the jumping leg of the locust is innervated by a "fast" fiber that, when stimulated, causes a maximum twitch providing power for the well-known jump of locusts and grasshoppers. Activity in a second, "slow" nerve fiber is responsible for gradual and partial shortening of the muscle and for the tonic contractions necessary in walking and standing. The inhibitory fibers found in crustacean muscles have not been unequivocally demonstrated in insects, although such an arrangement would seem to be most likely.

In these cases inhibition in one form or another would seem

to be an inevitable component of the motor mechanism for flexing and extending a limb. On the other hand, the significance of the inhibitory interactions postulated for the central control systems in the brains of mantids and crickets is not so obvious.

Central Inhibition. Most of the central inhibitory systems of insects have not yet been exposed to electrophysiological scrutiny. Electrophysiological methods revealed the endogenous nature of motor impulses connected with sexual movements in the mantis and cockroach (Chapter 10), but it has not yet been possible to detect activity in the inhibitory fibers postulated to originate in the subesophageal ganglion. The systems in the right and left halves of the brain of the mantis and cricket (Chapter 12) that, it was postulated, inhibit both one another and the excitatory locomotor centers in the central body and subesophageal ganglion remain completely unendowed with neural mechanisms.

The only inhibitory system in insects that has been adequately explored by electrophysiological methods is also one of the most curious. This is the interaction of photoreceptor cells in the ocellus with second-order ocellar nerve fibers (Chapter 9). Ruck[53] showed that excitation of the photoreceptor cells by light inhibits endogenous activity originating in the terminal processes of large axons that form the ocellar nerve. In continued darkness both the receptor cells and the ocellar nerve fibers may become endogenously active, so that the intermittent nerve signal transmitted to the brain is a compromise between the endogenous activity of the fibers transmitting it and the inhibition imposed by the endogenous activity of the photoreceptor cells (Fig. 41). Unfortunately no information is available on the behavioral role of the ocellus, but in any case it is difficult to see why the signaling of changes in light intensity should have required the evolution of such a complex signaling mechanism.

In another electrophysiological study Case[7] has shown that rhythmic ventilation of the tracheal system in the cockroach is determined by a pacemaker system located in the metathoracic and first abdominal ganglia. This system regulates and coordinates the arhythmic activity caused by carbon dioxide acting on neurons located in other parts of the nerve cord by regularly suppressing portions of their discharge. In this manner the respiratory muscles in different parts of the body are made to contract in concert.

These few examples are hardly a sufficient basis for generalizations, but in those that have been examined physiologically it has been shown that the output of a free-running or endogenously active group of nerve cells is patterned by inhibitory impulses originating upstream. In the case of the ocellus the upstream inhibitory neurons also tend to be endogenously active, and their output is in turn patterned, in this case by light stimulation. The systems in the brain and subesophageal ganglion controlling locomotion probably work in a similar manner, although this awaits electrophysiological examination. The control of endogenous activity by inhibition is common in other systems, for example vagus control of the vertebrate heart beat, and indeed it seems to be the only way in which such activity could be effectively regulated (page 154). Therefore, the ubiquity of inhibition in the systems we have examined in insects suggests that the elemental or uncontrolled state of many neurons and neuron systems in the nervous system is one of autonomous activity. A glance at the history of insects is relevant to this possibility.

The Origins of Insects. The phylum Arthropoda, which includes the Crustacea, Arachnida, Insecta, and several smaller classes, clearly shares a common ancestry with the phylum Annelida, or segmented worms. The development of a rigid and relatively watertight exoskeleton and jointed appendages enabled the arthropods to move out of the supporting mud inhabited by the soft-bodied worms, and eventually to occupy much of the water and land, as well as the air. The worms are generally regarded as the older and more conservative of the two phyla, and the present mode of existence of many marine forms is probably not very different from that of the ancestor common to both groups. The nature of this ancestor is of some interest, but because it was undoubtedly soft-bodied it has left no fossil traces beyond wormlike tracks found in some of the oldest sedimentary rocks.

Although a wasp and a worm are so different in appearance and way of life, their basic internal organization has many similarities. The nervous system, digestive tract, and heart have the same general relations, and in both forms the body is made up of a series of segments or metameres. In higher insects this metamerism has become much obscured by the fusion of some adjacent segments and by the loss of others, but it is still evident in the regular arrangement of the appendages and of the plates enclosing the abdomen, as well as in the sequence of ganglia that form the ventral nerve cord (Figs. 31, 56). The more marked segmentation found in less-specialized insects and in most larval forms leaves no doubt that this is a primitive characteristic.

In the most generalized annelids each body segment, excepting a few at the anterior and posterior ends, is similar to the rest. It has a complete set of muscles, excretory organs, circulatory organs, nerve centers, and often reproductive organs. Even though each segment has its organs coupled to those of the next, and the digestive tract runs through all, it is largely an autonomous and self-contained unit. This local autonomy is further demonstrated by the fact that a section consisting of a few segments removed from the middle portion of some worms not only is able to regenerate a new "head" and "tail," but also is capable of most of the peristaltic movements that constitute locomotion in this group.

In present-day annelids this autonomy is no longer complete. Among the several suprasegmental systems unifying some of the segmental activities, the best known is the system of giant fibers extending throughout the ventral nerve cord (Chapter 7). A noxious stimulus generates impulses in the giant fibers that suppress any current locomotor activity and bring about almost simultaneous contraction of the longitudinal muscles in all segments, so that the body of the worm is pulled back into its burrow.

It is probable that arthropods, and through them insects, evolved from a metameric creature in which the individual segments were even more independent and self-sustaining than this. The appearance of an exoskeleton, the fusion of some segments and the specialization of others with their appendages for specific functions, and the closer coupling of the body regions needed for a terrestrial existence, must have required the imposition of a series of suprasegmental command systems —the well-known process of cephalization. The point here is that if the original segmental activities were partly self-sustaining through local reflexes, and partly endogenous in the manner of the rhythmic peristalsis whereby some present-day annelids ventilate their burrows[67] then it would be natural for the suprasegmental control systems to operate with an inhibitory or negative sign. Serial as well as mutual feedback inhibitory systems of the sort postulated above for insects could have arisen in the following way. After a certain segment had come to suppress, or at least to regulate with a negative sign, the activities of one or more segments at another level, it might have become subjected in its turn to similar inhibitory control from another, superior source.

This would account for the prevalence of endogenous activity in isolated insect ganglia (Chapter 9), and make it easier to see the significance of the endogenous motor activity released in the terminal abdominal ganglion of the mantis and cockroach after removal of the subesophageal ganglion (Chapter 10). It was pointed out that this mechanism seems to have survival value in the male mantis, since fertilization is still possible, and sometimes even more probable, after the cannibalistic attack by the female. On the other hand, a similar neural mechanism in the male cockroach could not be found to have any significance in adaptive behavior. Perhaps transection of the nerve cord in both insects merely reveals the primordial local automatisms of a segmental mechanism. In the mantis this primordial automatism has secondarily achieved adaptive significance and become reinforced through the sexual selection that stems from the cannibalistic tendencies of the female. This may also be true about the tendencies of many female insects to oviposit after destruction of the head and thorax by a predator. In the male cockroach the same neural mechanism has been subjected to no such selective reinforcement, and remains merely as a partially organized relic of the past.

One of the great contributions made by modern ethology has been to emphasize that animal behavior has a phyletic history that closely parallels evolution of the structures so familiar to comparative anatomists. Darwin recognized this more than 100 years ago, but there was little firm evidence at that time. In the past 20 or 30 years there have been a number of studies in which a specific behavior pattern together with the structures and actions that release it have been compared in series of variously related animal species and genera. This material now constitutes the substantial and rapidly growing literature of comparative ethology. It has been suggested that the absence of a fossil record of behavior makes any conclusions about specific and generic relationships a matter of guesswork if they are based on the comparison of current behavior patterns. This stricture applies equally well to most of the conclusions based on comparative anatomy, particularly those dealing with the history of soft organs such as the circulatory system, also to most of systematics at the generic and specific level. All responsible zoologists recognize the dangers of extrapolating from the present into the past, but in the absence of a paleontology of, say, the heart or kidney, or of a paleoethology, there seems to be no way of establishing animal relationships other than the comparative method that is the basis of most zoology.

The suggestion made here about the origin of the system of inhibition *cum* local autonomy found in insect nervous systems is of this kind. It is not presented as an explanation, but rather as a *point d'appui* for future experiments. Perhaps the acquisition of species survival value by the endogenous mechanism in the terminal ganglion of the mantis and possibly other insects is unique, and the system persists in other insects and other parts of the nervous system merely because it happens to be the functional ground plan. On the other hand, endogenous mechanisms may play a more direct part as the basis for appetitive behavior (Chapter 9), also enabling the animal to extract itself from the possible, although not very probable, cul-de-sac of minimal or zero stimulation.

Neural Parsimony. It is estimated that there are more than 1,000,000 living species of insects. The great majority have a body weight of only a small fraction of an ounce. There are only about 30,000 living vertebrate species, some weighing as much as 40 tons. The diversity of form and the limited size of insects are usually attributed to the principles of their body architecture—a subject beyond the scope of this book.

In spite of this limit to their body size, insects have some nerve cells and axons that are larger in diameter than our own. From this it follows that insect central nervous systems must consist of fewer neurons. No actual count has ever been made, but estimates suggest that the insect brain and ventral nerve cord contains fewer neurons than one vertebrate eye and optic nerve.

Yet insects must compete diversely for their survival against larger animals more copiously equipped. They must see, smell, taste, hear, and feel. They must fly, jump, swim, crawl, run, and walk. They must sense as acutely and act as rapidly as their predators and competitors, yet this must be done with only a fraction of the number of their nerve cells. This has imposed on insect evolution an economy—I have called it parsimony[34] —of nerve cells. Each nerve cell must be capable of handling a maximum amount of information, and there are probably few "stand-bys" or alternative pathways. This principle of parsimony is most obvious in connection with the nervous mechanisms of evasive behavior. For instance, the great jumping muscle of the grasshopper has only a few motor axons, while our analogous muscles are supplied with tens of thousands. The sacrifice of informational detail for speed of operation was discussed at length when describing the giant interneuron system of the cockroach (Chapter 7). The two receptor cells of the moth ear supply an example of this principle on the sensory side.

This frugal use of available mechanisms has made possible most of the experiments described above. Present methods permit estimates of the total information transfer only when a small number of channels are surveyed. Furthermore, the parsimonious arrangements of insects are to me more attractive to contemplate and more intellectually satisfying than the prodigality exemplified by vertebrate nervous systems. The very unscientific term "ingenious" comes to my mind when thinking about how insects have solved their communications problems. This is what has drawn me to the study of insects' lives and nervous systems and to the writing of this book.

References

1. E. D. Adrian, "The activity of the nervous system of a caterpillar," *J. Physiol. 70,* 34–35 (1930); "Potential changes in the isolated nervous system of *Dytiscus marginalis,*" *J. Physiol. 72,* 132 (1931).

2. Biological Clocks, *Cold Spring Harbor Symposia on Quantitative Biology 25* (1960).

3. A. D. Blest, T. S. Collett, and J. D. Pye, "The generation of ultrasonic signals by a new-world arctiid moth," *Proc. Roy. Soc. Lond. Ser. B 158,* 194–207 (1963).

4. H. F. Blum, *Time's Arrow and Evolution* (Princeton University Press, Princeton, N.J. 1955).

5. M. A. B. Brazier, *The Electrical Activity of the Nervous System* (Macmillan, New York, ed. 2, 1960).

6. F. Brink, D. W. Bronk, and M. G. Larabee, "Chemical excitation of nerve," *Ann. N. Y. Acad. Sci. 47,* 457–485 (1947).

7. J. F. Case, "Organization of the cockroach respiratory center," *Biol. Bull. 121,* 385 (1961); *Amer. Zoologist 1,* 440 (1962).

8. M. J. Cohen, "The function of receptors in the statocyst of the lobster, *Homarus americanus,*" *J. Physiol. 130,* 9–34 (1955).

9. D. C. Dunning and K. D. Roeder, "Moth sounds and insect-catching behavior of bats," *Science 147,* 173–174 (1965).

10. J. C. Eccles, *The Physiology of Nerve Cells* (Johns Hopkins Press, Baltimore, 1957). *The Physiology of Synapses* (Academic Press, New York, 1964).

11. F. Eggers, papers on the anatomy and function of tympanal organs in moths, *Zool. Jahrb. Abt. Anat. 41,* 273–376 (1919); *Z.*

vergleich. Physiol. 2, 297–314 (1925); *Schrift. Naturw. Ver. Schleswig-Holstein 17,* 325–333 (1926); *Die Stiftführenden Sinnesorgane* (Zoologische Bausteine, Berlin, 1928).

12. K. von Frisch, *Bees: their Vision, Chemical Senses, and Language* (Cornell University Press, Ithaca, N.Y., 1950); K. von Frisch and M. Lindauer, "The 'language' and orientation of the honeybee," *Ann. Rev. Entomol. 1,* 45–58 (1956).

13. D. R. Griffin, *Listening in the Dark* (Yale University Press, New Haven, Conn., 1958); *Echoes of Bats and Men* (Doubleday, New York, 1959).

14. D. R. Griffin, F. A. Webster, and C. R. Michael, "The echolocation of flying insects by bats," *Animal Behaviour 8,* 141–154 (1960).

15. D. R. Griffin, J. Friend, and F. Webster, "Target discrimination by the echolocation of bats," *J. Exper. Zool. 158,* 155–168 (1965).

16. E. von Holst, papers on central nervous function and spontaneous activity in earthworms and fish, *Zool. Jahrb. 51,* 4 (1932); *53,* 1 (1933); *Z. vergleich. Physiol. 21,* 5 (1934); *23,* 2 (1936); *24,* (1937); *26,* 3 (1938).

17. E. von Holst and U. von Saint Paul, "Vom Wirkungsgefüge der Triebe," *Naturwissenschaften 18,* 409–422 (1960); "Electrically controlled behavior," *Sci. Amer. 206,* 50–59 (1962).

18. G. Hoyle, *Comparative Physiology of the Nervous Control of Muscular Contraction* (Cambridge University Press, Cambridge, England, 1957).

19. F. Huber, papers on central nervous control of locomotion, grooming, singing, and courtship in crickets, *Z. Tierpsychol. 12,* 12–47 (1955); *Zool. Anz., suppl. 23,* 248–269 (1959); *Z. vergleich. Physiol. 44,* 60–132 (1960).

20. P. Karlson and A. Butenandt, "Pheromones (ectohormones) in insects," *Ann. Rev. Entomol. 4,* 39–58 (1959).

21. K. S. Lashley, "In search of the engram," *Symp. Soc. Exper. Biol. 4,* 454–482 (1950).

22. M. Lindauer, *Communication among Social Bees* (Harvard University Press, Cambridge, Mass., 1961).

23. K. Z. Lorenz, "The comparative method of studying innate behaviour patterns," *Symp. Soc. Exper. Biol. 4,* 221–268 (1950).

24. M. M'Cracken, "The egglaying apparatus in the silkworm

(*Bombyx mori*) as a reflex apparatus," *J. Comp. Neurol. 17,* 262 (1907).

25. N. Milburn, E. A. Weiant, and K. D. Roeder, "The release of efferent nerve activity in the roach, *Periplaneta americana,* by extracts of the corpus cardiacum," *Biol. Bull. 118,* 111–119 (1960).

26. N. Milburn and K. D. Roeder, "Control of efferent activity in the cockroach terminal abdominal ganglion by extracts of corpora cardiaca," *J. Gen. Comp. Endocrinol, 2,* 70–76 (1962).

27. H. Mittelstaedt, "Prey capture in mantids," in B. T. Scheer, ed., *Recent Advances in Invertebrate Physiology* (University of Oregon Publications, Eugene, Ore., 1957), pp. 51–71; "Control systems of orientation in insects," *Ann. Rev. Entomol. 7,* 177–198 (1962).

28. S. Ozbas and E. S. Hodgson, "Action of insect neurosecretion upon central nervous system *in vitro,* and upon behavior," *Proc. Nat. Acad. Sci. 44,* 825–830 (1958).

29. R. S. Payne, K. D. Roeder, and J. Wallman, "Directional sensitivity of the ears of noctuid moths," *J. Exper. Biol. 44,* 17–31 (1966).

30. R. J. Pumphrey, "Hearing," *Symp. Soc. Exper. Biol. 4,* 3–18 (1950).

31. S. Rilling, H. Mittelstaedt, and K. D. Roeder, "Prey recognition in the praying mantis," *Behaviour 14,* 164–184 (1959).

32. K. D. Roeder, "An experimental analysis of the sexual behavior of the praying mantis," *Biol. Bull. 69,* 203–220 (1935).

33. K. D. Roeder, "The control of tonus and locomotor activity in the praying mantis (*Mantis religiosa L.*)," *J. Exper. Zool. 76,* 353–374 (1937).

34. K. D. Roeder, "Organization of the ascending giant fiber system in the cockroach (*Periplaneta americana L.*)," *J. Exper. Zool. 108,* 243–262 (1948).

35. K. D. Roeder, "The effect of potassium and calcium on the nervous system of the cockroach," *J. Cell. Comp. Physiol. 31,* 327–338 (1948).

36. K. D. Roeder, chapters 17 and 18 in K. D. Roeder, ed., *Insect Physiology* (Wiley, New York, 1953).

37. K, D. Roeder, "Spontaneous activity and behavior," *Sci. Monthly 80,* 362–370 (1955).

38. K. D. Roeder, "A physiological approach to the relation between prey and predator," *Smithsonian Inst. Misc. Collections 137,* 287–306 (1959).

39. K. D. Roeder, "The predatory and display strikes of the praying mantis," *Med. Biol. Illust. 10,* 172–178 (1960).

40. K. D. Roeder, "The behaviour of free flying moths in the presence of artificial ultrasonic pulses," *Animal Behaviour 10,* 300–304 (1962).

41. K. D. Roeder, "Echoes of ultrasonic pulses from flying moths," *Biological Bulletin 124,* 200–210 (1963).

42. K. D. Roeder, "Aspects of the noctuid tympanic nerve response having significance in the avoidance of bats," *J. Insect Physiol. 10,* 529–546 (1964).

43. K. D. Roeder, "Acoustic sensitivity of the noctuid tympanic organ and its range for the cries of bats," *J. Insect Physiol. 12,* 843–859 (1966).

44. K. D. Roeder, "Interneurons of the thoracic nerve cord activated by tympanic nerve fibers in noctuid moths," *J. Insect Physiol. 12,* 1227–1244 (1966); "Auditory system of noctuid moths," *Science 154,* 1515–1521 (1966).

45. K. D. Roeder, "A differential anemometer for measuring the turning tendency of insects in stationary flight," *Science 153,* 1634–1636 (1966); "Turning tendency of moths exposed to ultrasound while in stationary flight," *J. Insect Physiol. 13,* 890–923 (1967).

46. K. D. Roeder, L. Tozian, and E. A. Weiant, "Endogenous nerve activity and behaviour in the mantis and cockroach," *J. Insect Physiol. 4,* 45–62 (1960).

47. K. D. Roeder and A. E. Treat, "Ultrasonic reception by the tympanic organs of noctuid moths," *J. Exper. Zool. 134,* 127–158 (1957).

48. K. D. Roeder and A. E. Treat, "The acoustic detection of bats by moths," paper presented at the XI International Entomological Congress, Vienna, 1960.

49. K. D. Roeder and A. E. Treat, "The detection of bat cries by moths," in W. Rosenblith, ed., *Sensory Communication* (Technology Press, Cambridge, Mass., 1961).

50. K. D. Roeder and A. E. Treat, "The detection and evasion of bats by moths," *Amer. Scientist 49,* 135–148 (1961).

51. L. M. Roth and E. R. Willis, "The reproduction of cockroaches," *Smithsonian Inst. Misc. Collections 122,* 1–49 (1954).

52. T. C. Ruch, H. D. Patton, J. W. Woodbury, and A. L. Towe, *Neurophysiology* (Saunders, Philadelphia, 1961).

53. P. Ruck, "Electrophysiology of the insect dorsal ocellus," *J. Gen. Physiol. 44,* 605–627, 629–639, 641–657 (1961).

54. F. Schaller and C. Timm, "Das Hörvermögen der Nachtschmetterlinge," *Z. vergleich. Physiol. 32,* 468–481, (1950).

55. B. Scharrer, "Neurosecretion XI. The effects of nerve section on the *intercerebralis-cardiacum-allatum* system of the insect *Leucophaea maderae," Biological Bulletin 102,* 261–272 (1952).

56. J. Ten Cate, "Physiologie der Gangliensysteme der Wirbellosen," *Ergeb. Physiol. 33,* 137–336 (1931).

57. N. Tinbergen, *The Study of Instinct* (Oxford University Press, London, 1951).

58. W. H. Thorpe, "Comparative psychology," *Ann. Rev. Psychol. 12,* 27–50 (1961); W. H. Thorpe and O. L. Zangwill, eds., *Current Problems in Animal Behaviour* (Cambridge University Press, Cambridge, England, 1961).

59. A. E. Treat, "The response to sound of certain Lepidoptera," *Ann. Entomol. Soc. Amer. 48,* 272–284 (1955).

60. A. E. Treat, "The reaction time of noctuid moths to ultrasonic stimulation," *J. New York Entomol. Soc. 54,* 165–171 (1956).

61. A. E. Treat and K. D. Roeder, "A nervous element of unknown function in the tympanic organs of noctuid moths," *J. Insect Physiol. 3,* 262–270 (1959).

62. D. M. Vowles, "The structure and connexions of the corpora pedunculata in bees and ants," *Quart. J. Microscop. Sci. 96,* 239–255 (1955).

63. D. M. Vowles, "Neural mechanisms in insect behaviour," in W. H. Thorpe and O. L. Zangwill, eds., *Current Problems in Animal Behaviour* (Cambridge University Press, Cambridge, England, 1961).

64. A. Watanabe and T. H. Bullock, "Modulation of activity of one neuron by subthreshold slow potentials in another in lobster cardiac ganglion," *J. Gen. Physiol. 43,* 1031–1045 (1960).

65. F. Webster, "Bat-type signals and some implications," in E. M. Bennett, J. Degan, and J. Spiegel, eds., *Human Factors in Technology* (McGraw-Hill, New York, 1962); F. Webster and D. R. Griffin, "The role of the flight membranes in insect capture by bats," *Animal Behaviour, 10,* 332–340 (1962).

66. E. A. Weiant, "Control of spontaneous efferent activity in certain efferent nerve fibers from the metathoracic ganglion of the

cockroach," in A. E. R. Downe and W. G. Friend, eds., *Proceedings of the X International Entomological Congress* (Mortimer, Ottawa, 1956), vol. 2, pp. 81–82.

67. G. P. Wells, "Spontaneous activity cycles in polychaete worms," *Symp. Soc. Exper. Biol. 4,* 127–142 (1950).

68. V. B. Wigglesworth, *The Control of Growth and Form* (Cornell University Press, Ithaca, N.Y., 1959).

69. D. M. Wilson, "The central nervous control of flight in a locust," *J. Exper. Biol. 38,* 471–490 (1961).

70. E. O. Wilson, "Chemical communication among workers of the fire ant *Solenopsis saevissima* (Fr. Smith)," *Animal Behaviour 10,* 134–164 (1962).

71. R. W. G. Wyckoff and J. Z. Young, "The motor neurone surface," *Proc. Roy. Soc. (London) Ser. B. 144,* 440–450 (1956).

Index